D0851480

LIVING EARTH

TITLES OF RELATED INTEREST

Igneous petrogenesis
B.M. Wison

Invertebrate palaeontology and evolution (2nd edition)
E.N.K. Clarkson

Microfossils
M.D. Brasier

A natural history of Nautilus
P. Ward

Paleopalynology
A. Traverse

Quaternary paleoclimatology
R. Bradley

Rates of evolution
P.S.W. Campbell & M.F. Day

Trace fossils
R.G. Bromley

Vertebrate palaeontology
M. Benton

LIVING EARTH

A SHORT HISTORY OF LIFE AND ITS HOME

E.G. Nisbet
Department of Geological Sciences
University of Saskatchewan

HarperCollins*Academic*
An imprint of HarperCollins*Publishers*

Published by
HarperCollins*Academic*

77–85 Fulham Palace Road
Hammersmith
London W6 8JB
UK

10 East 53rd Street
New York, NY 10022
USA

First published in 1991

British Library Cataloguing in Publication Data

Nisbet, E. G. (Euan George)
Living Earth.
I. Title
575
ISBN 0–04–445855–X
ISBN 0–04–445856–8

Library of Congress Cataloguing-in-Publication Data

Nisbet, E. G.
Living Earth: a short history of life and its home / E.G. Nisbet.
p. cm.
Includes bibliographical references and index.
ISBN 0–04–445855–X.
ISBN 0–04–445856–8
1. Life–Origin. 2. Geobiology. 3. Historical geology.
I. Title.
QH325.N56 1991
577–dc20
90–49552
CIP

Typeset in 11 on 13 Palatino by
Phoenix Photosetting, Chatham
Printed in Great Britain by
Bell and Bain

To P.A.N.

Contents

Preface

Life has shaped the Earth, and the Earth has moulded the history of life. That history, the co-evolution of our ancestors and their home, has much to teach us about our place on the planet today. We are part of the fabric of the biosphere. As we change that fabric we would be wise to understand how our home was built.

Our planet is neither a hotel nor a colony. It is not a place which life briefly inhabits during a transient occupation. Instead, it is our home, designed by the deeds of our ancestors and suited to our own needs. The history of life on Earth is held in the geological record, which is composed of the rocks, water and air that are available for study on the planet's surface. These rocks, the oceans and the atmosphere are not simply stores of information for the excitement of fossil hunters and geochemists, or resources to exploit without thought. Their creation and continued existence form an integral part of the development and management of the Earth as the home of life.

The purpose of this book is to tell the story of life on Earth; to show how life has influenced and structured the planet, and how the planet, in turn, has shaped life; and to show how we, as human inhabitants of the Garden of Eden, rely on the rest of the Garden for our own existence. We were born monarchs of the Garden, but a monarch can fall. We may consider that we rule nature, but our rule, like that of the Queen of England, depends on the community. The traditional role of the monarch is that of a shepherd, guiding and protecting the flock, and depending on it. If the community is misdirected or destroyed, the monarch suffers too, or falls.

The word 'economy' is rooted in the Greek concept of stewardship of the home, but modern economic theory has little to say about stewardship. It assumes resources are to be developed, without thought to the stewardship of the planet. In contrast, our understanding of geological history is based on James Hutton's notion of the 'oeconomy of the world'. It is fine to build a bigger and better home, or even a palace, but our palace should be that of the responsible ruler, not the fortress of a tyrant, or our wealth will become corruption and, eventually, our palace will become ruins.

I teach a large introductory class in Earth history, attended by students who are not majoring in science. I have been struck by the deep interest that many students show in the history of our environment. Standard introductory texts in geology do not address this. They are filled with raw facts and explanations of the workings of the Earth, but there is rarely any sense of how life and the planet evolved together as a system. The attention is given to the component parts, not to the whole.

In this book, I have attempted to show, with analogies, how the interrelationship between life and the planet grew, to become the modern biosphere. I have written the book with introductory students in mind, in the hope that it can be used by them as a supplement to a standard text. The book is also addressed to the general reader, who perhaps took a basic science course years ago and now would like to learn more. The story is exciting, complex, interesting. It is important that we know our history: as we attempt to manage the biosphere today, we can succeed only if we know our past. I have tried to convey the excitement, to simplify and make accessible the complexity, and to demonstrate the interest and importance of the story. My hope is that the reader will be drawn in, to realize that geology is not just a collection of fossils, minerals and rocks. Instead, it is the story of our home, how it was built, and how our ancestors lived in it and shaped it. It is the family story of humanity.

Inevitably in describing so complex a subject I have in places stepped close to the edge of controversy. Many things are not understood. Geology is a living science and there is still much to learn. Where possible, I have outlined both sides of the controversy. Occasionally, I have followed the advice given to me by a scientist who played a major part in the discovery of plate tectonics: 'Be bold, and hope your judgement is right!'. The more important themes are developed in stages, with reiteration, throughout the book. Some subjects, such as natural selection and the role of sex, I have glided over gently, for which I apologize, but the texture of this book and the fabric of my knowledge is too light to sustain detailed exposition.

This book is only a brief introduction to the history of life on earth. Suitable further reading is listed at the end of each chapter, for those who wish to look deeper. The university libraries and bookshops are filled with orthodox books of geology. The interested reader will, I hope, be persuaded to go on, to discover the monsters that lurk ahead. My hope is that I can encourage students to search out the real literature of science, the scientific journals, to encounter the thrill of discovery in the pages of nature. But I must warn them too: be sceptical. Our knowledge of the biosphere remains poor and partial.

I should like to thank S.H. Alexander, E. Belt, W. Braun, W.G.E. Caldwell, R. Carroll, S. Conway Morris, C. Earle, P.H. Fowler, M. Hynes, S. and B. Ingham, D. Greenwood, L. Margulis, L. Nelson, C. Newton and W.A.S. Sarjeant, all of whom read the manuscript in detail or made many strong comments, which were much valued. The errors that remain are, of course, my fault not theirs, especially where I have gone against their advice! I would also like to thank all those who commented on parts of the manuscript, helping me to remove errors or infelicity of style. My special thanks to Roger Jones, for encouraging and carefully criticizing the book, to Andy Oppenheimer, Caroline Hannan, Kim Dean and Angaïs Scott and in particular to Angie Heppner, who put it together.

Acknowledgements

The Royal Tyrrell Museum of Palaeontology, in Drumheller, Alberta, Canada, director Emlyn Koster, is gratefully thanked for kindly allowing me to use illustrations by its artist Vladimir Krb, to whom I express my appreciation. I also thank Ted Irving for allowing me to use his palaeocontinental maps (reprinted by permission of Kluwer Academic Publishers); Paul Copper, Harry Whittington and Simon Conway Morris for their diagrams; Bill Compston for a photograph of ancient zircon; John Sclater and Chris Scotese for maps (Figure 1.15 by permission of Elsevier Science Publishers); Mike Bickle, Tony Martin, Lynn Margulis, Hank Williams and Geoff Brown for illustrations; and most especially Mary Fowler, who made the book possible (Figures 1.11, 1.14 and 1.18 by permission of Cambridge University Press). I wish to thank the museum of the Geology Department, University of Saskatchewan for many of the line drawings. Figures 8.3, 8.4a, and 8.5–7 are from a publication of Energy, Mines and Resources, Canada, reproduced with the permission of the Minister of Supply and Services, Canada. The extract from *Soko Risina Musoro (Shona)* by H.W. Chitepo translated and edited by Hazel Carter (OUP for the School of African and Oriental Studies, 1958) was reprinted by permission of Oxford University Press. The extract from *The Lama* by Ogden Nash © 1931 was reprinted by permission of Little, Brown and Company and André Deutsch Ltd.

Finally, I wish to express my appreciation of the late M.H. Lister, who gave me her transcription of the letters of A.G. Bain, and encouraged my interest in geology.

'He has made everything beautiful in its time;
also he has put eternity into our hearts, yet so that no-one
can understand what God has done from the beginning to the end'

Ecclesiastes 3.11

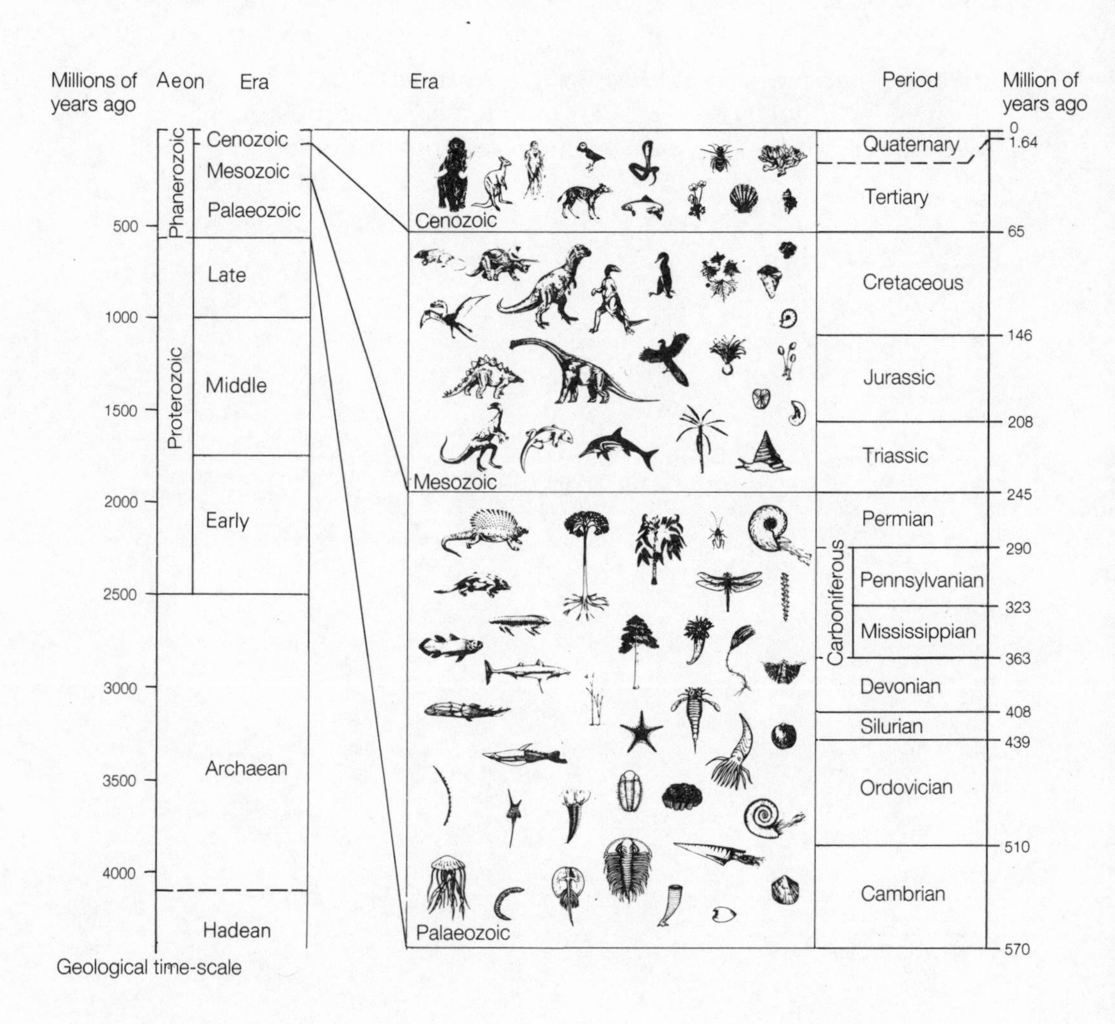

The geological time scale

Life probably began at the beginning of the Archaean Eon, roughly four billion years ago. Towards the end of the Proterozoic Eon, complex multicelled organisms evolved. Hard parts first appeared at the beginning of the Cambrian Period, and make up most of the fossil record. Human beings have evolved within the past few million years (courtesy of Department of Geology, University of Saskatchewan).

PART I

In the abyss of time

1
In the beginning

THE EARLIEST EARTH

The Earth is old, and in its fabric is written a subtle and complex history. The oldest material ever discovered on Earth consists of a few crystals of a mineral called zircon. These crystals, found in Western Australia, are about 4.3 billion years old. Many of the elements that make up the Earth are roughly 4.6 billion years old, and the planet probably formed soon after that, although much of the hydrogen in our water may be as old as the universe. For the first 300 million years that Earth existed there is no history recorded in our terrestrial rocks,

Figure 1.1 A zircon crystal from Western Australia, 4.15 billion years old. This crystal, and others like it that range up to nearly 4.3 billion years old, formed a few million years after the accretion of the Earth. Scale bar is 100 microns long (courtesy of W. Compston).

although the Moon, meteorites and comets preserve earlier material. Some meteorites even preserve traces of events that took place before the formation of the Solar System.

The Solar System probably began as an interstellar cloud of dust and gas. There is much debate about how the Sun and planets were formed, but the most likely explanation was first put forward over two hundred years ago by the philosopher Immanuel Kant and the mathematician Pierre Laplace, who suggested that the collapse of a cloud of interstellar material, falling together under its own gravitation, could lead to the formation of a central star and orbiting bodies. In the modern development of this hypothesis, the collapse of the cloud may have been triggered by the shock wave from the explosion of a nearby star into a supernova. The evidence for this is that some meteorites contain what appear to be products of the rapid radioactive decay of unstable atoms, such as plutonium or radioactive aluminium, formed in one or more stellar explosions that must have taken place shortly before the formation of the Solar System. By studying the daughter atoms from the decay of these short-lived atoms we can deduce that the supernova explosion took place not long before the formation of the planets. Comets may include some very primitive fragments of the interstellar cloud; some day we may be able to bring back samples of them.

As the cloud contracted, it turned into a spinning disc of material (a nebula), with the Sun forming at the centre. The cloud heated while it contracted and began to rotate, as material fell inwards and collided. Eventually the young proto-Sun became large enough and hot enough to sustain nuclear reactions, while small bodies of collided material orbited it. These small bodies, known as planetesimals, may have included bodies as large as tens of kilometres across. The biggest planetesimals, with more gravitational pull, grew faster than the smaller objects. They swept up debris from their orbital region and drew in smaller bodies, to become the planets. In the outer, cool part of the solar nebula, the cloud was rich in hydrogen, helium, water and carbon compounds, that eventually collected as the gas-giant planets. Closer to the Sun, the higher temperatures meant that the low-boiling-point compounds became gases and were lost, and the result was the inner rocky, or terrestrial, planets: Mars, Earth, Venus and Mercury are made of rock around an iron-rich core.

Accretion of the planets probably took place very rapidly. A large planetesimal, once formed, would gravitationally attract other material in a similar orbit around the Sun. By collision with this material the planetesimal would grow bigger and more attractive until it had swept out all the smaller material in its orbit, and had become a planet in birth. As material fell into the growing planet, the collisions would be very energetic, and the planet would heat up rapidly.

The last collisions in the main phase of accretion may have been gigantic. It is possible that near the end of the Earth's growth period there was a huge collision between the young Earth and another planetary body about as large as Mars. This impact would have thrown out a ring of molten rock around the Earth. The ring then condensed to form the Moon. Our Earth–Moon system rotates as a double planet. The impact may have given us the rotation that has evolved to our present day-length, and it may have helped to tilt the Earth's axis, to give us winter and summer. Perhaps it was a critical event in making the Earth habitable.

We can measure the age of the Solar System by dating meteorites and also by looking at the history of lead produced on the Earth by the radioactive decay of uranium. Some meteorites may be fragments that were produced very early in history of the Solar System, and their age can be determined by measuring the composition of radioactive elements and their daughter products. Similarly, the age of the Earth can be calculated by analysing the history of the various different types of lead atoms in the Earth's crust. The Earth and meteorites are 4.55 to 4.6 billion years old, implying that the collapse of the dust cloud and the formation of the planets took place around that time.

The young Earth was hot. The heat came both from the gravitational energy that was released as accreting material fell into the planet, and also from radioactive decay of uranium, thorium, potassium, aluminium and plutonium in the material. Today, the Earth remains radioactive, but much less so. All the radioactive aluminium and plutonium atoms are gone. Aluminium cooking pots, for instance, are not 'hot' in the radioactive sense. There are only a few atoms of natural plutonium left. Radioactive heat in the modern Earth is generated by the remaining naturally radioactive elements: the potassium, uranium and thorium that were originally made in a supernova. Over the planet's history geological processes have concentrated these elements in the Earth's outer layers, where they produce heat.

The rate of radioactive heat generation during the birth of the Earth would have been many times greater than today. The planet was so hot that much of the inner part of the early Earth may have been molten. Even today, although most of the Earth is now solid, part of the heat which produces volcanoes and earthquakes, is derived from the primeval store of heat. This store of heat, gained during the catastrophic falling-in of the accretion and from early radioactive decay, has helped to drive the geological evolution of the planet.

Some geologists refer to this pregeological time as the **Hadean** Eon, the time of hell-fire. The main phase of accretion may have taken only a few million years, but it left an organized Solar System around a central Sun, comprising planets that were chemically differentiated. The Earth, by the end of the Hadean, probably had a central iron-rich **core**, a **mantle** made of magnesian silicates and a **crust** of cooled lava and perhaps granitic rocks. As the planet accreted, gas was driven off from the hot partly-molten interior. Much of the gas must have been lost to space, driven off by the particles and radiation streaming from the young Sun. But, as accretion ended, an atmosphere was left on the planet's surface. Studies of the radioactive decay history of some of the rare gases, such as xenon, show that they were present in our atmosphere very early on, and it is probable that the young Earth had a substantial early atmosphere dominated by water (mostly condensed as the ocean), carbon gases and nitrogen compounds. Impacts continued after the main phase of accretion – meteorites still hit us today, and heavy bombardment of the Earth and Moon probably continued until about 3.5 billion years ago. This bombardment produced the scars on the face of the full Moon.

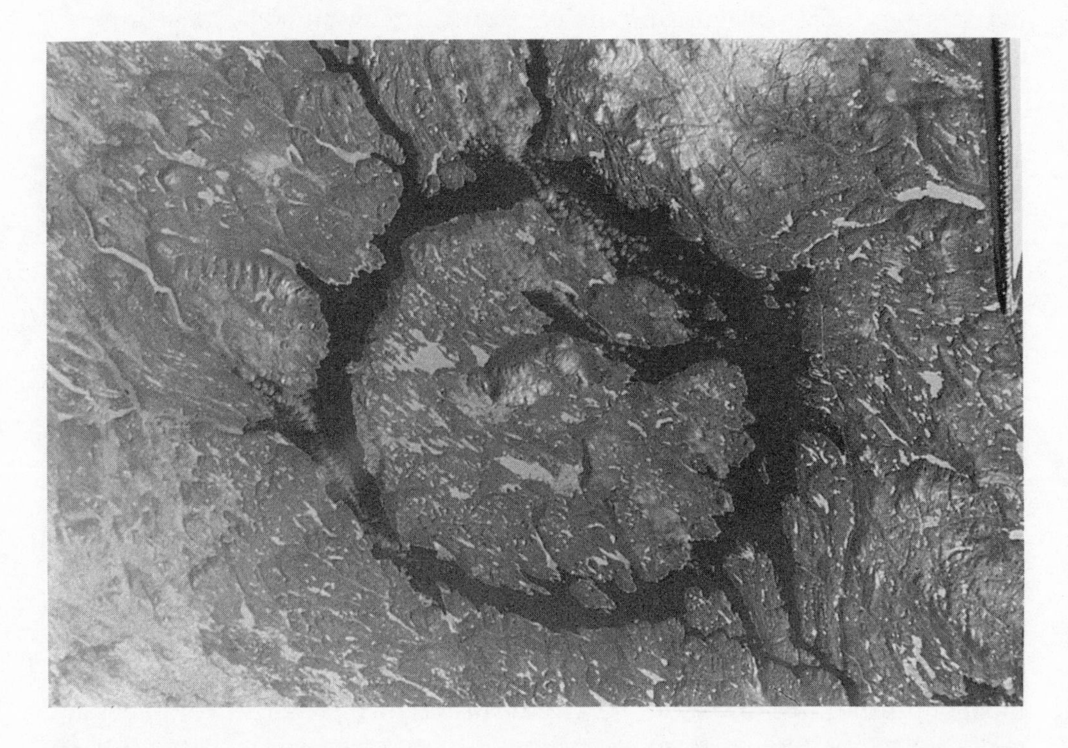

Figure 1.2 Manicouagan impact site, Quebec province, Canada. The diameter of the site is 70–100 km (45–60 miles), and the impact was relatively recent in geological time, about 210 million years ago (courtesy of NASA).

By about 4.4 billion years ago, the Earth had been formed, its character determined. At this stage there was nothing exceptional about the Earth. It was an inner, rocky planet probably very similar to the young Venus, with a hot interior and a surface rich in water. It orbited a young Sun. Yet Venus evolved to the modern inferno, a dry ball of rock with a dead atmosphere, while the Earth remains a wet and living planet.

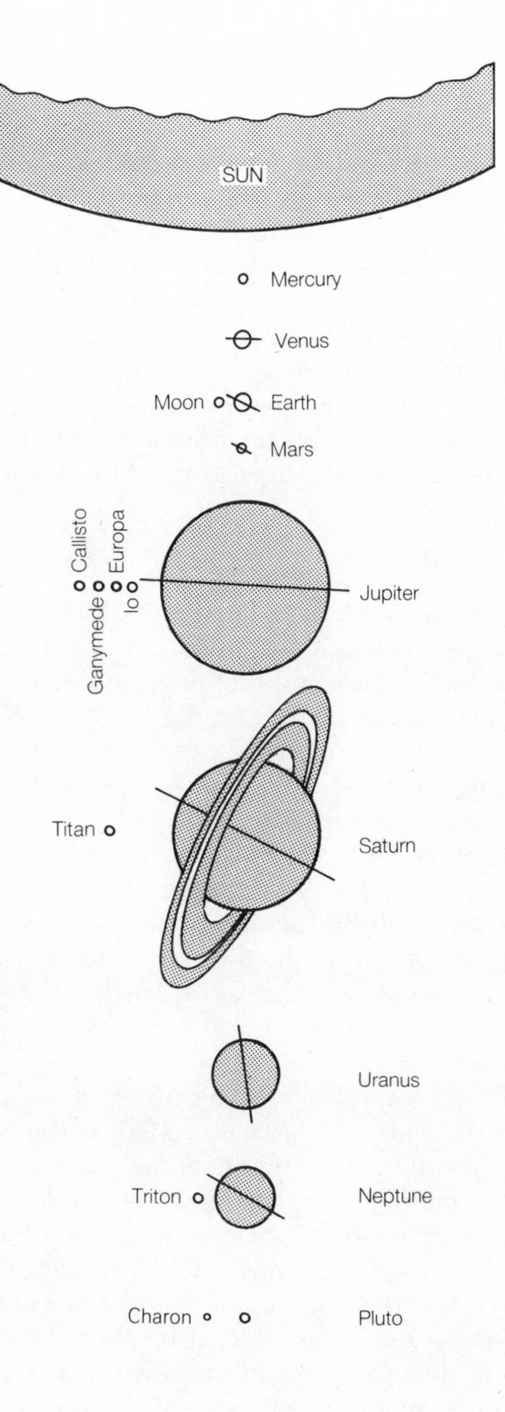

Figure 1.3 The Sun and the planets, showing the relative sizes of the Sun, planets and bigger moons. Distances between the bodies are not to scale. (From Brown, G.C. & A.E. Mussett 1981 *The inaccessible Earth*. London: Allen & Unwin.)

THE WATER CRISIS

The oldest proper rocks (as opposed to single crystals) known are about 4.0 to 3.8 billion years old. Rocks of this age are known from northern Canada, Antarctica, and from the Isua area in West Greenland. The interval of about five hundred million years between the origin of the Earth and the formation of the first rocks in the geological record is about the same length of time as from the beginning of the Cambrian Period, when the main fossil record began, to the present. During this unrecorded time, both the Earth and Venus probably passed through a water crisis. Venus emerged dry and hot, while Earth became the home of life. It is worth examining this water crisis in some detail, because the presence of liquid water is perhaps the most important factor in our inorganic heritage to have shaped the development of the modern Earth.

Scylla, or the greenhouse trap

The discovery that Venus may have been wet deserves to rank amongst the nicest of the happy accidents that amuse the scientist. There is rivalry in science, just as everywhere else, and Venus had become known as the Russian planet, whereas Mars was an all-American affair. This superpower rivalry is, for once, good-natured, co-operative and creative. The Soviet Venera spacecraft made some remarkable discoveries: the surface of Venus is very hot (500°C; 900°F); the atmosphere is dry and dominated by carbon dioxide, at surface pressures of about 90 Earth atmospheres (or 90 bars); the rocks seem to be fragments of lava, possibly basaltic. Venus has been mapped by radar in some detail. The results show high areas, low areas, terrains looking like a giant example of a crumpled-up carpet, huge craters and enormous lava flows. But there seems to be no twofold division as on Earth into high continental areas and low oceanic areas, and Venus does not have a slightly lopsided centre of mass, as the Earth does.

When, eventually, an American NASA craft did descend through the atmosphere, one of the experiments was designed to investigate the composition of the air of Venus. Unfortunately, the inlets became clogged at an altitude of about 50 km by sulphuric acid droplets from the clouds that cover Venus, and the experiment appeared to be a failure in its attempt to study the planet's lower air. Here came the happy accident: much later, after the initial dismay had died away, it was realized that the clogging of the inlet had accidentally created an instrument that was very sensitive in hydrogen analysis. There is very little hydrogen on Venus, but the clogged instrument behaved like a hydrogen enrichment chamber. From this lucky accident, it was discovered that the atmosphere of Venus has a high deuterium : hydrogen ratio. In other words, some of its hydrogen is heavy: the stuff used in some nuclear reactors and that occurs in small quantities in Earth's oceans, making heavy water.

Figure 1.4 Venus. The clouds include sulphuric acid droplets, in an atmosphere of CO_2. The planet is slightly smaller than Earth (courtesy of NASA).

If the interpretation is correct, this rather obscure measurement is perhaps the most interesting single result to come from our exploration of the Solar System. It suggests that some process has selectively removed enormous amounts of light hydrogen from the atmosphere of Venus, disproportionately leaving behind the heavier hydrogen, deuterium, so that in the little hydrogen that remains the ratio of deuterium to light hydrogen is very high. The most probable explanation for this is that Venus once had oceans of water (H_2O), and that the hydrogen was removed from moisture in the upper atmosphere by solar ultraviolet radiation, which detached the H from the H_2O, allowing the light hydrogen to be lost to space. The deuterium, which is twice as heavy, was lost to space less easily than the light ordinary hydrogen. Oxygen was left behind. Calculations imply that Venus may have had oceans as deep as a kilometre or more: in other words, the planet was once like the Earth, although much closer to the Sun and therefore collecting more solar heat. This interpretation is still controversial and there are other explanations , but if it is true its significance is fascinating: a wet Venus could have lived like the Earth. Instead, it became dry and died. Why?

What seems to have happened is that a 'greenhouse' environment developed on Venus. A domestic garden glasshouse keeps plants warm by trapping hot air, stopping it from rising and carrying heat away. In an 'ideal' greenhouse, if the air is still, there is another reason that the inside is warm: energy enters through the glass as light of short wavelength from the hot Sun. This short-wavelength – visible – light, (light we can see) passes through the glass. The interior of the greenhouse warms, and radiates heat back. Radiated heat is infra-red, or long-wavelength light: we can feel it, but we cannot see it. It cannot pass through glass – to it, the glass is a wall. The heat is trapped. Thus the inside of the 'ideal' greenhouse warms up, and heat can be lost only by conduction through the glass, not by radiation. An atmosphere works in much the same way as an ideal greenhouse. Gases such as carbon dioxide act like glass. They allow the Sun's radiation to enter, but block radiation leaving the planet. The trapped heat warms up the atmosphere and surface until enough infra-red radiation (heat) is given off to maintain the balance.

In the early history of Venus, a greenhouse developed, so that the temperature of the water on the planet's surface was high enough to at least produce a very moist upper atmosphere, or even to boil much of the sea into the atmosphere. The first condition is known as a moist greenhouse, with a surface temperature perhaps as low as 40–95°C (100–200°F); even if the original temperature was relatively low, the initial wet conditions were followed by a catastrophe in which the seas gave off water until they boiled dry. If there were large quantities of moisture in the high atmosphere, then hydrogen would have been rapidly lost to space, removed by the effects of the ultraviolet-rich radiation of the Sun, and it has been estimated that as much water as there is in the modern Earth's oceans could have been lost from Venus in roughly 500 million years, which is geologically not a long time. Venus today, with its surface temperature of 500°C, has what is called 'runaway greenhouse'. The planet is dry; life cannot exist.

Earth is much further from the Sun than is Venus: on Earth the oceans remain. We do not know how near the Earth came to sharing the fate of Venus. The modern Earth atmosphere is characterized by a feature known as a **cold-trap**; in other words, it is coldest not at the top or bottom, but in the middle. On average, much of the surface of the world is at a pleasant 10–25°C (50–80°F). If we ascend in an aircraft which has a flight-information screen, the temperature can be seen to drop far below freezing at 35 000 feet (about 10 km). This is near the cold trap. At much higher altitudes the temperature rises again. At the outermost fringes of the atmosphere, the temperature is very high.

Moisture lost by evaporation from the oceans cannot, in general, rise above the cold trap. At this level the air is saturated with water. Cold air can carry very little moisture. Above this level the air cannot carry more water in it than the small amount saturating the cold trap. The moisture reaches the cold part of the atmosphere, condenses and eventually falls down again as rain, and air travellers often see a marvellous flat tablecloth of cloud below them and clear sky above. The existence of the cold trap means that only small amounts of

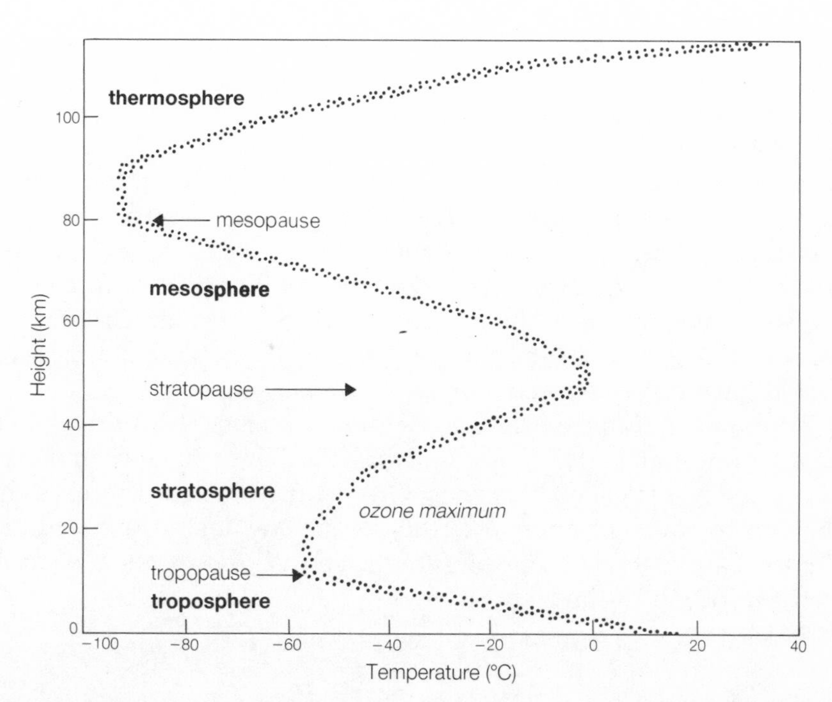

Figure 1.5 The average temperature of the Earth's modern atmosphere, as it varies with height. Near the ground, the air is warm. At the tropopause, which is the top of the troposphere, the air is cold, and very little water can pass upwards above this level: instead it condenses out and falls back to the surface. Higher up is the stratosphere, which contains air that is much more settled than in the turbulent troposphere. Above this are the mesosphere and the thermosphere.

Figure 1.6 The Earth: Africa, Antarctica and the southern oceans (courtesy of NASA).

water reach the high upper atmosphere, which is for the most part dry, and therefore poor in hydrogen in H_2O. As a result, very little hydrogen is today lost to space from the Earth, because so little hydrogen is available at the top of the atmosphere to be lost.

Our oceans are safe, at least from the Sun. It is possible that this has not always been true, and that a moist greenhouse did once exist on Earth, in which water could reach the higher atmosphere. If so, hydrogen would have been lost, and oxygen would have remained. However, Earth being further from the Sun than is Venus, the loss of hydrogen would have been much slower than on Venus, where any early cold trap would have been much warmer than on Earth and would have allowed much more water to pass upwards and eventually to space. By about four billion years ago, Venus was probably utterly dehydrated, having lost almost all of its hydrogen, while Earth may have been partly dehydrated but still had substantial oceans. One estimate is that perhaps as much as one-third of the present volume of the oceans was lost in this way from the Earth. If the Earth had lost some of its hydrogen, abundant oxygen would have been left behind, in the air.

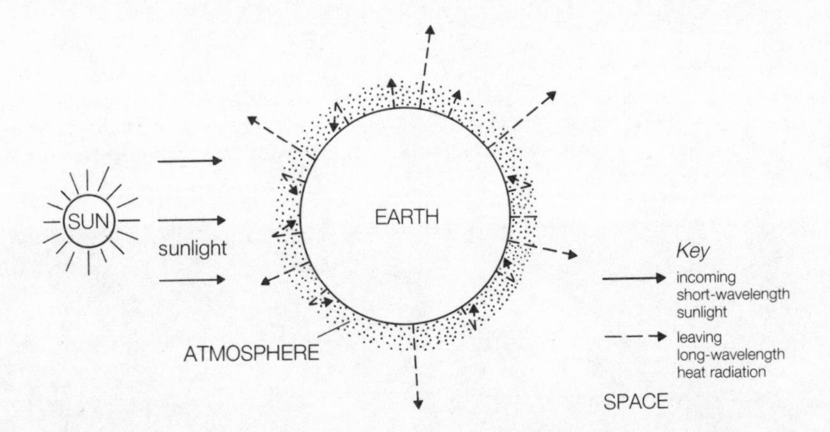

Figure 1.7 The greenhouse effect. Incoming radiation (short wavelength) from the Sun enters the atmosphere and is absorbed in the air, by clouds and on the ground. Heat radiation leaving the Earth, which is of longer wavelength, is absorbed in part by greenhouse gases and water vapour in the air, which act as a blanket around the planet. To maintain a balance, with incoming radiation equalled by outgoing and thus a steady temperature, the Earth is warmer than it would be if the greenhouse gases were not present. Without water, carbon dioxide and methane, the Earth's surface temperature would be, on average, roughly −18°C (0°F). In fact, it is about +15°C (59°F) on average: the difference is due to the greenhouse effect of the natural atmosphere.

All this, of course, is speculation. We really do not know what happened. But whatever took place, the events in those earliest years of the atmosphere and ocean were critical to the later development of the planets. Since then, carbon dioxide has continued to bubble out from the interiors of Venus and Earth. Both Earth and Venus have large and similar amounts of carbon on or near the planetary surface (this is called the surface inventory) but whereas on Venus the carbon dioxide is present as gas in the air, on Earth most of it is present in limestones. Which brings us on to the next great problem.

Charybdis, or the Icebox Trap

If the standard model of stellar evolution is correct, the young Sun was faint, although it produced more than it does today of the ultraviolet radiation capable of dislodging hydrogen from the Earth. Four and a half billion years ago, much less heat – only two-thirds to three-quarters of the current amount – reached the Earth. In this lies what is called the solar luminosity paradox. The young Sun was cooler. But the modern Earth was until recently on the verge of glaciation and we have only just emerged from a period when a significant fraction of the northern hemisphere was covered in ice. Today we still have an entire continent, Antarctica, under ice. Imagine, then, the situation if the Sun suddenly cooled to the state it was in billions of years ago – the planet would suffer catastrophic global glaciation and the seas would all freeze. Once such a catastrophe had taken place, there would be no escape, because the white ice would reflect the solar radiation. Even if the Sun became much hotter again, it would be unable to melt an ice-covered planet. If the planet is covered by ice, a situation explored by Kurt Vonnegut in his book 'Cats Cradle', there is an end to life. The planet dies. It cannot recover. Earth has had many glaciations, but it has always escaped this catastrophe.

Yet if the modern Earth, with a strong warm Sun, has recently been through a glaciation, how did it avoid glaciation over four billion years ago, when the Sun was weak? Obviously some special factor allowed a liquid ocean to exist then. Water-lain sediments are known from over the whole four billion year time span recorded in the rocks. The temperature has stayed fairly steady, with fluctuations, but no catastrophe. The only answer to this solar luminosity paradox is that on the young Earth there was a substantial greenhouse contribution to the surface temperature. Although the effective temperature (the temperature of the planet in the absence of a greenhouse contribution) was well below freezing, four billion years ago the greenhouse effect of water vapour, together with gases such as carbon dioxide or methane, must have been sufficiently large to allow a surface on which liquid water, not ice, filled the oceans. But if the Earth was warm four billion years ago, why are the oceans not boiling today? Why is Earth not now an inferno?

The explanation of this paradox is that, as the Earth grew older and the Sun grew warmer, the greenhouse contribution to the surface temperature was steadily reduced. Today the greenhouse effect adds an extra 33°C (59°F) to the Earth's surface temperature, raising the temperature from −18°C (0°F), which it would be in the absence of air, to about +15°C (+59°F). In contrast, on modern Venus the greenhouse atmosphere has warmed the surface by several hundred

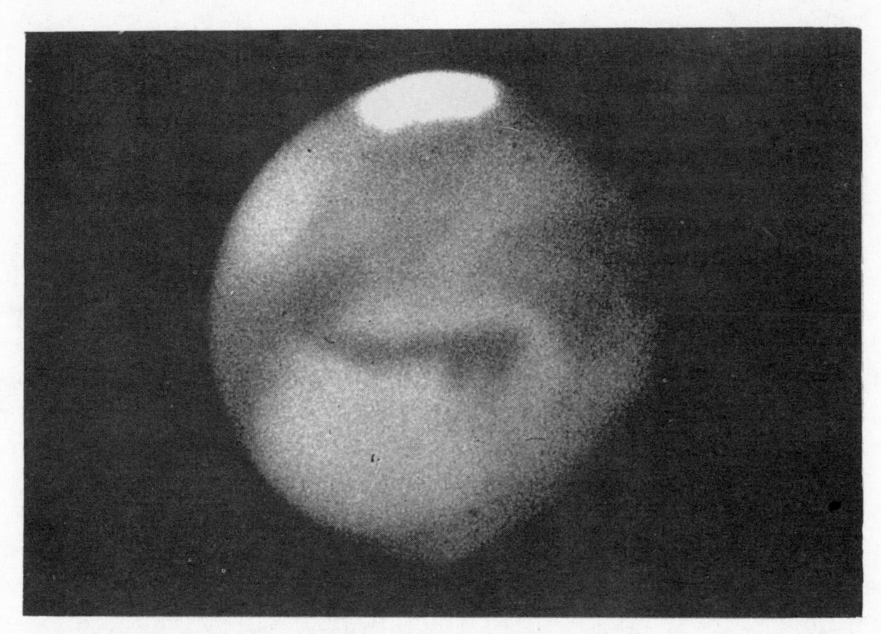

Figure 1.8 Mars. The polar ice cap includes dirty water ice. The planet is roughly half the diameter, and one tenth the mass of Earth (courtesy of NASA).

Figure 1.9 The surface of Mars. On a hot summer's day, near the equator, the temperature rises close to 0°C (32°F) (courtesy of NASA).

degrees centigrade. The warming of the early Earth must have been an atmospheric process, not heat arising inside the Earth, since we can calculate that the contribution of heat from inside the Earth to the surface temperature must have been trivial except in the very earliest part of the Earth's history. From four billion years ago until today, internal heat has driven the geological activity of the planet but, when compared to solar radiation, it is miniscule in its effects on the atmosphere.

A PLANET FIT FOR LIFE

Like Ulysses, the Earth found the middle course between Scylla and Charybdis. The early disaster of a catastrophic greenhouse was avoided, but not at the cost of an equally catastrophic glaciation. The Earth's surface was warm enough to maintain oceans of liquid water four billion years ago, yet those oceans are not boiling today. The planet was obviously lucky, being in a nearly circular orbit at such a distance from the Sun that the effective temperature in the absence of air was only slightly below freezing, but the explanation of the paradox demands more than the mere chance of an orbit. Some process seems to have compensated for the gradual warming of the Sun by steadily decreasing the initial greenhouse contribution and, in so doing, maintaining liquid oceans and a place fit for habitation.

What can we say about the surface environment of the Earth when the planet had stabilized after its accretion, say 4.3 billion years ago? It is indisputable that an atmosphere existed (the rare gases show that), and it is very likely that deep oceans existed. Beyond that, little is certain. Most models of the early atmosphere suggest that the air was composed mainly of carbon dioxide, associated with nitrogen, together with water vapour. There is little likelihood that methane and ammonia were present at the earliest stage. Even if these two compounds had been present, they would probably have been destroyed rapidly: ammonia, for example, may have been destroyed in less than 50 years by the effect of light from the Sun.

We do not know if free oxygen was present. Most scientists think it came later, but some have suggested that it was in the air at a very early stage. One observation supporting this opinion is that some old rocks do seem to have been created in oxidizing conditions. Furthermore, there is some physical evidence and logic supporting the inference that a short-lived moist greenhouse did develop on Earth. Perhaps a kilometre or even more of the original depth of the ocean was lost, as hydrogen was knocked out into space. Loss of hydrogen to space from water would have left oxygen behind, producing an oxidized atmosphere. This perhaps even oxidized part of the planet's interior. However, although there is disagreement about how much oxygen was present, most scientists agree that the young atmosphere contained carbon dioxide and some nitrogen and water vapour, over a planetary surface largely covered by deep oceans.

By the end of the Hadean, about four billion years ago, the Earth had settled into its present state – a rocky planet with an active surface and a differentiated interior. Its physical constitution has evolved somewhat since then, but the essential fabric of the planet was complete.

THE CONSTITUTION OF THE SOLID EARTH: PLATE TECTONICS

Since Isaac Newton's time, earth scientists have gradually unravelled most of the mysteries of the Earth's physical operation. We now know what the Earth is made of, how it moves, and why geological events such as earthquakes and volcanic eruptions occur. Although there are still many gaps in our knowledge, the broad outlines of the Earth's constitution and functioning are well understood.

The Earth weighs just under 6×10^{27} (six, followed by 27 zeros) grammes, or about 6×10^{21} tonnes. This is slightly heavier than Venus, and much more than Mercury, Mars or the Moon, but much less than Jupiter or Saturn. The average density is about 5.5 times that of water, or about double the density of typical surface rocks such as granite. The implication of this high density is that the deep interior of the planet is dense and iron-rich, even allowing for the high density that results where rocks are squashed by compression deep in the Earth.

The circumference of the Earth was first measured by the ancient Greeks: it is about 40 000 km (25 000 miles). To Captain Cook, the first scientist to see the Earth as a whole, this was an enormous distance – to us, it is a two-day flight, or a ninety-minute space orbit. The planet is small.

The interior of the Earth has been mapped in detail, using earthquakes in much the same way that doctors use X-rays. Around the turn of the century John Milne, an early seismologist who worked in Japan, discovered that the outer layer of the Earth, the **crust**, is quite different from the layer beneath. The boundary at the base of the crust was discovered by Andrya Mohorovičić, a Croatian seismologist, after whom it is now called the '**Moho**'. We now know that most of the continental crust is about 35–40 km (20–25 miles) thick.

Some of the continental crust is made of sedimentary rocks, such as sandstones and limestones, laid down on the surface. Other parts of the crust are made of igneous rocks derived from molten magma intruded into the crust or erupted as lava by volcanoes. Yet other rocks, such as marble, first formed as sedimentary or igneous rock and were then altered by changes in pressure or temperature, to become metamorphic rocks. The oceanic crust is different and much thinner. It is made up of igneous rock overlain by varying thicknesses of sediment, and is typically 7–10 km (about 4–6 miles) thick.

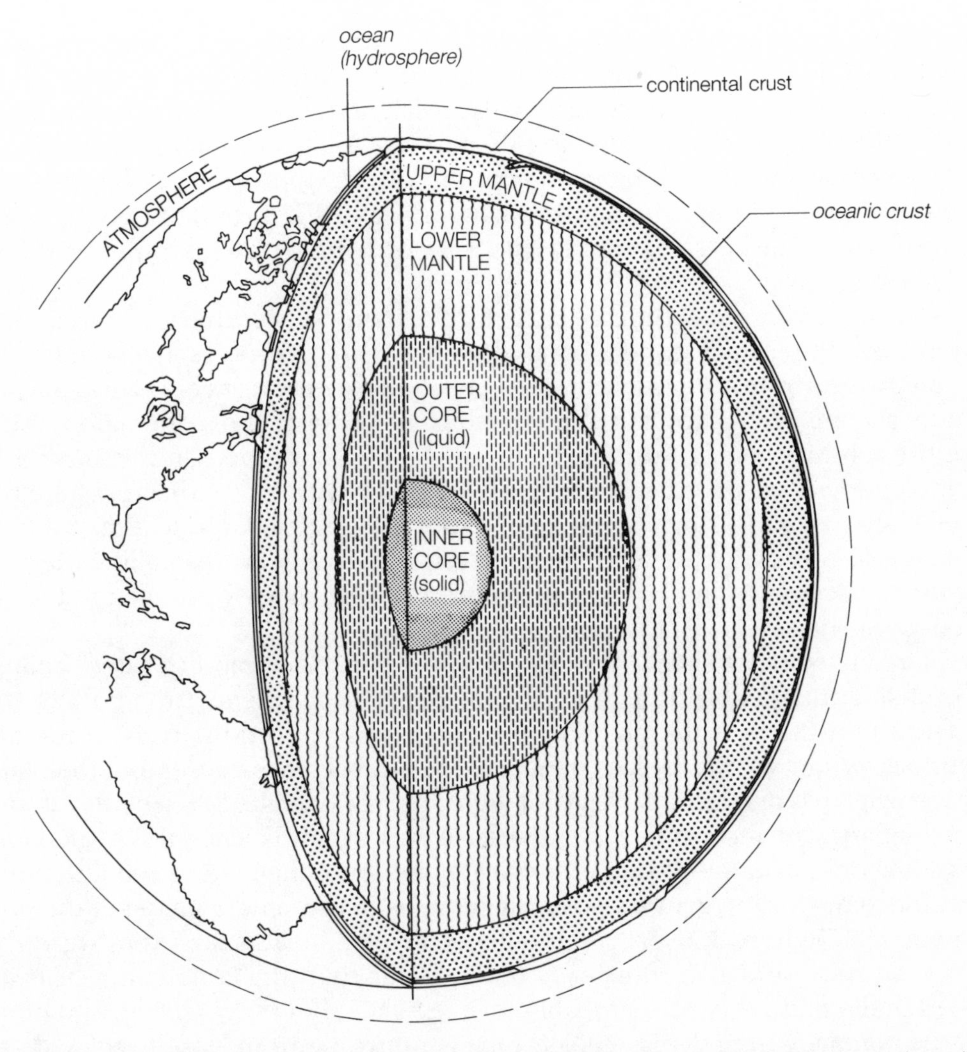

Figure 1.10 The constitution of the Earth. The oceanic crust is roughly 8 km (5 miles) thick, the continental crust is about 35 km (22 miles) thick. At the base of the crust is the Moho. Below the crust is the Upper Mantle, down to about 670 km (415 miles), and then the Lower Mantle, to 2900 km (1800 miles). The Outer Core (2900–4980 km; 1800–3100 miles) is liquid; the Inner Core is solid (courtesy of Department of Geology, University of Saskatchewan).

The deeper structure of the Earth has also been mapped out by using earthquake waves. Below the Moho, to a depth of 2900 km (about 1800 miles), is the **mantle**, divided into the **upper mantle** (roughly the top 700 km, 415 miles) and the **lower mantle**. The mantle is mainly made of magnesium–iron–silicate minerals, such as olivine. The innermost region of the Earth, below 2900 km (1800 miles) is the **core**, that is itself divided into the fluid outer core, and the solid inner core. The core is dense, and probably rich in iron. It is the source of the Earth's magnetic field.

The modern Earth is an active planet. It has volcanoes and earthquakes, and its history is one of moving continents and the making and destruction of oceanic basins. These are the surface expressions of loss of heat from the interior of the planet to the surface and then to space. This heat loss is the driving force behind the surface behaviour of the planet, a process called **plate tectonics**. Tectonics simply means the carpentry of the Earth, and the word **plate** is used because the Earth's surface is made up of distinct plate-like sectors, rather like a soccer ball.

Heat is transferred within the Earth by two methods: conduction and convection. Near the surface, conduction is dominant. In this process, heat is transferred upwards from atom to atom, each hot atom exciting its neighbour, but each atom staying in its place. For a simple household example of conduction, think of the conduction of heat through the metal of a hot pipe. On the Earth as a whole, however, pure conduction mostly occurs near the surface, where the rocks are cool and strong. Much more important on a global scale is the transfer of heat by convection, whereby hot material physically moves, taking the heat with it, just as in winter hot air rises from an air duct or above a radiator, to heat a house. The air carries the heat.

A good analogy to heat transfer within the Earth is found in a pot of boiling Scottish oatmeal porridge. Roughly speaking, heat enters at the bottom (by conduction through the metal of the pot) and hot porridge rises to the top surface, where it cools by loss of heat to the air, becomes denser and then falls back when it is displaced by new hot, light, rising porridge. Much of the interior of the Earth convects in this way, because rock expands when it is heated and becomes less dense. Although the silicate interior of the Earth is solid, the rock is able to move slowly in a viscous fashion, as in the simple model of the porridge. Hot, light rock rises. Over millions of years, heat is carried upwards by moving rock, and this transfer is much faster than the conduction of heat. Within the Earth there are probably regions where hot rock is rising, and other large regions where dense, cooled rock is returning from close to the surface back to the interior.

But this picture of the way in which heat moves in the Earth is too simple. It may approximate what goes on in a dry planet, but it is not a true model of what goes on in either a pot of porridge or the Earth. Both systems are wet. In the porridge, water at the bottom of the pot boils, turns to steam and bubbles up. By doing so it drives the circulation; density changes in the porridge itself are minor factors in comparison. In the Earth, things are more complex, but the role of water is crucial. The interior of the Earth normally melts only at a very high temperature, and the deeper the rock is buried (i.e. the higher the pressure), the higher the temperature at which the rock will melt. If a rock rises, in a convecting system, it undergoes a pressure drop but does not necessarily lose heat; as a result, it may partially melt, separating into a solid residue and a liquid magma. The magma will not usually have the same composition as the original

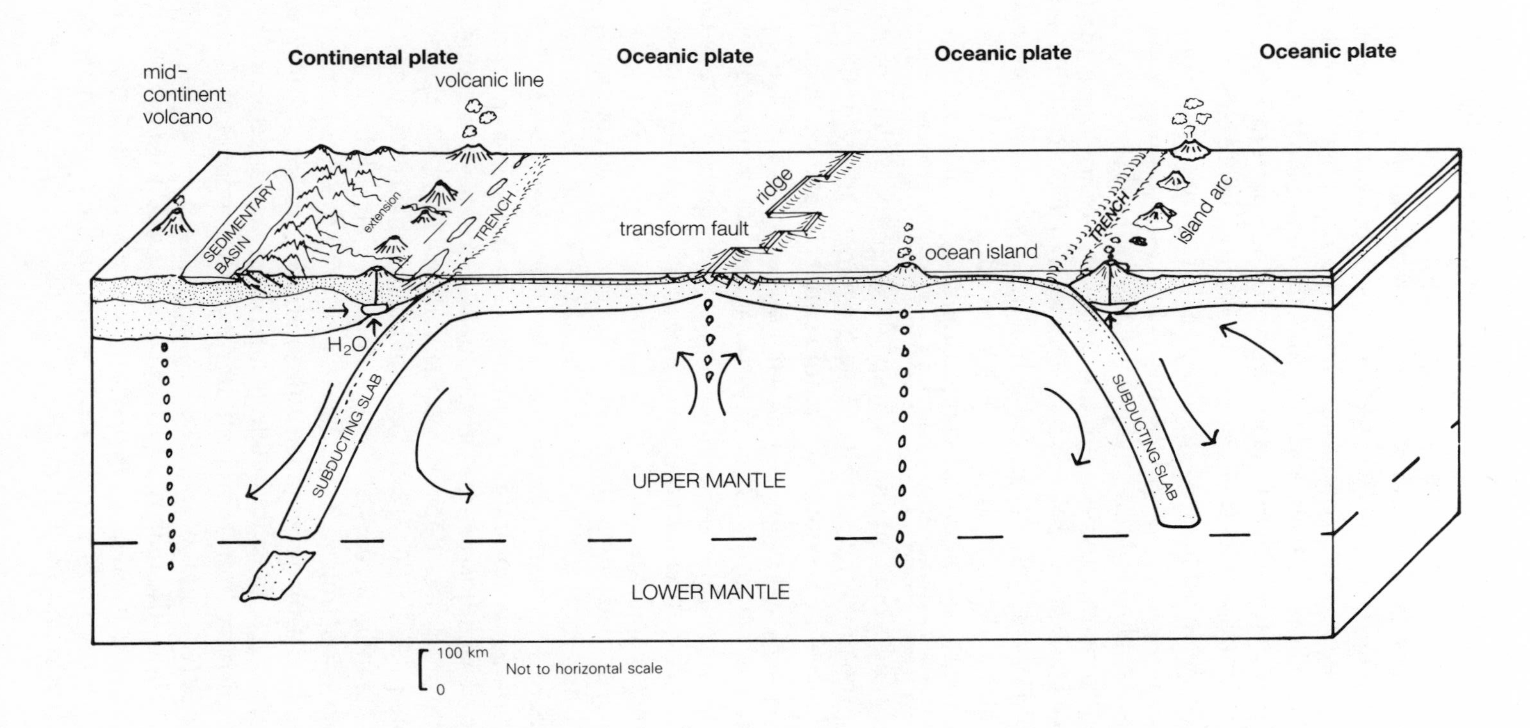

Figure 1.11 A cartoon of the cycle of plate tectonics, not to scale. Partial melting takes place under mid-ocean ridges, where new ocean floor is created. The development of new mid-ocean ridges can split continents apart. Elsewhere, old ocean floor falls back into the interior at subduction zones. The release of water above subducted ocean floor causes the overlying material partly to melt. The melt rises and goes through a complex set of processes under the overlying crust. Eventually, lava is erupted at volcanoes above the subduction zones, or granites are intruded into the crust (from Fowler, C.M.R 1990 *The solid Earth*. Cambridge: Cambridge University Press).

rock, and it will also usually (though not always) be much lighter than the original rock. The light melt will rise as it forms, taking heat with it rapidly to higher levels in the Earth, and it may even eventually erupt as a lava. Water can play a critical role in this process, because wet rocks usually melt extensively at much lower temperatures than dry rocks. If water is present on a planet the whole system will operate under cooler conditions than on a dry planet, since the water in the rock allows melt to form and take heat up to the surface.

On the Earth the process of heat loss takes a very definite pattern; volcanoes do not just bubble at random. Much of the Earth's internal heat is lost via the **mid-ocean ridges**. At these ridges new hot magma comes up from the upper mantle, as the ocean floor splits apart by a few centimetres a year. The new magma is injected along a line splitting the pre-existing ocean floor, and is also erupted on top of the older ocean floor. In this way, new ocean floor is made; each year a few cubic kilometres of new ocean floor are created. Rapid cooling follows the formation of the new ocean floor, mostly as seawater penetrates into the new, hot material. Vast volumes of water flow through the porous hot lava. On the modern Earth, a volume of water roughly equivalent to the total volume of the oceans passes through the new lava about once every ten million years, which is a very short time, geologically speaking. In the distant past the rate of cycling was probably much greater. This movement of water through the new, hot crust at the ridge is known as **hydrothermal circulation**. It has two effects: it cools the lava, so that hydrothermal cooling accounts for a large part of the heat loss from the interior; and it helps to control the chemistry of the seawater, because through this circulation chemical components are either leached out of the rock and enter into the sea or else precipitate out of seawater into veins in the new ocean floor.

The new ocean floor spreads away from the mid-ocean ridge as yet more lava is erupted at the axis, and cools further, producing a strong cold slab of material on the Earth's surface. This slab is known as the **lithosphere**, which is the mechanically strong outer part of the Earth, where heat transfer is by conduction, not convection. The lithospheric slab gradually thickens as it cools; the cooling and hence the thickness depends on the square root of the age of the slab, so that lithosphere that is 36 million years old is roughly twice as thick as lithosphere that is nine million years old. Eventually the slab falls back into the interior again, at an oceanic trench. This process by which the ocean floor falls back into the interior of the planet is called **subduction**. The term comes from Julius Caesar, who used it to describe pulling ships onto the English beaches, and was used in English by John Milton, to describe the removal of Adam's spare rib, in *Paradise lost*.

At the subduction zones the water, which was introduced into the oceanic crust by the hydrothermal circulation, is driven off the descending slab. This water rises. The driven-off fluid causes partial melting in the overlying material, which is injected with enormous quantities of water, together with carbon

Figure 1.12 Ngauruhoe volcano, North Island, New Zealand. This volcano, above a subduction zone, is frequently in eruption: a recent flow may be seen on its slopes, to the right of the picture. White patches are snow.

Figure 1.13 Ocean island volcanism, Kilauea volcano, Hawaii. This lava is less blocky and flows more easily than the lava of Ngauruhoe volcano that was shown in Figure 1.12. Oceanic island volcanoes such as those on Hawaii, which occur in the interiors of plates and not at plate boundaries, are not directly part of plate tectonics but are fed with magma from below the plates.

dioxide, chlorine and elements such as rubidium. The melt produced is light, and it rises. Many complex melting and crystallization processes occur as the melt rises, but derived liquids eventually reach the surface. The surface expression of the melting is a volcano. Mt St Helens, most of the great volcanoes of the Andes, Japan and Indonesia, Kamchatka in easternmost Siberia, and the volcanoes of the West Indies are all examples of this process.

These volcanoes characteristically produce a magma known as andesite (more than 55% silica), in contrast to the mid-ocean ridge lavas, which are basalt (usually around 49–50% silica). The andesitic lavas, together with similar magmas that are not erupted but emplaced inside the Earth, make up an important part of the continental crust, which is much richer in silica than the oceanic crust as a result.

The whole process is highly organized, and the Earth's surface can be divided into a small number of lithospheric caps. These are the *plates*, and each plate moves as a unit. As a result, most geological activity, such as earthquakes or volcanism, occurs at or near the boundaries between the plates. Very little happens in the interiors of plates, with the exception of some within-plate volcanic activity, for example at Hawaii. There are three main sorts of boundaries between plates: **constructive**, or extensional, as at the mid-ocean ridges where new lithosphere is created; **destructive**, or lines of consumption, where one plate is consumed down a subduction zone; and **conservative** boundaries, lines where plates slide against each other, as along the San Andreas Fault in California. The whole process of plate movement is subject to a rule of spherical geometry called Euler's theorem, which allows plate motions to be calculated. From this mathematical relationship it is possible to work out the motion of any part of the Earth's surface, given a small amount of data mostly from mid-ocean ridges. Recently, these calculated motions have been confirmed by direct measurements of movement between plates, using satellite and radio-telescope techniques. From the motions it is possible to predict what sort of earthquake is likely in any particular place.

Set into these moving plates are the great light rafts we call continents. Oceanic lithosphere may grow dense and fall back into the interior, but continents are made of lighter material and are not, for the most part, swallowed back into the planet. The continents are made of enormously complex aggregations of material, including sedimentary accumulations (sandstone, limestone, and the like) on top, volcanic rock, granitic intrusions and a deep continental crust of heated, recrystallized, metamorphosed rock that is very approximately of granitic composition, much richer in silica than the ocean floor. Although some of the components are dense, such as uranium, in bulk the continents are made of material that is lighter than the ocean floor, and so the continents form the high part of the Earth's relief. Furthermore, the Earth is very slightly lopsided because one hemisphere is full of light, high-standing continents, while the other side is mostly the Pacific Ocean. On Mars the

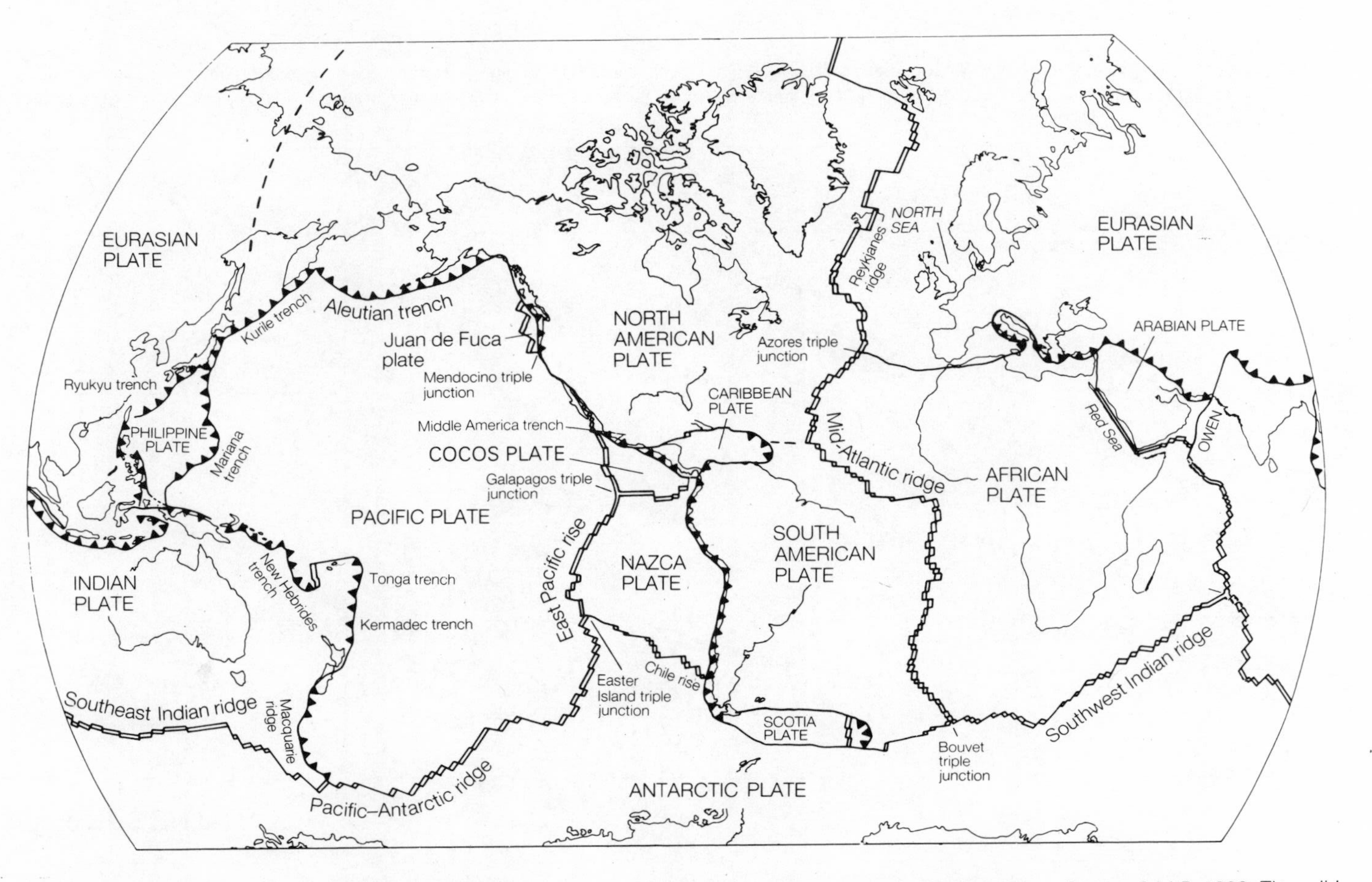

Figure 1.14 Map of the world's plate system, showing the major plates, ridges, trenches and transform faults (from Fowler, C.M.R. 1990. *The solid Earth*. Cambridge: Cambridge University Press).

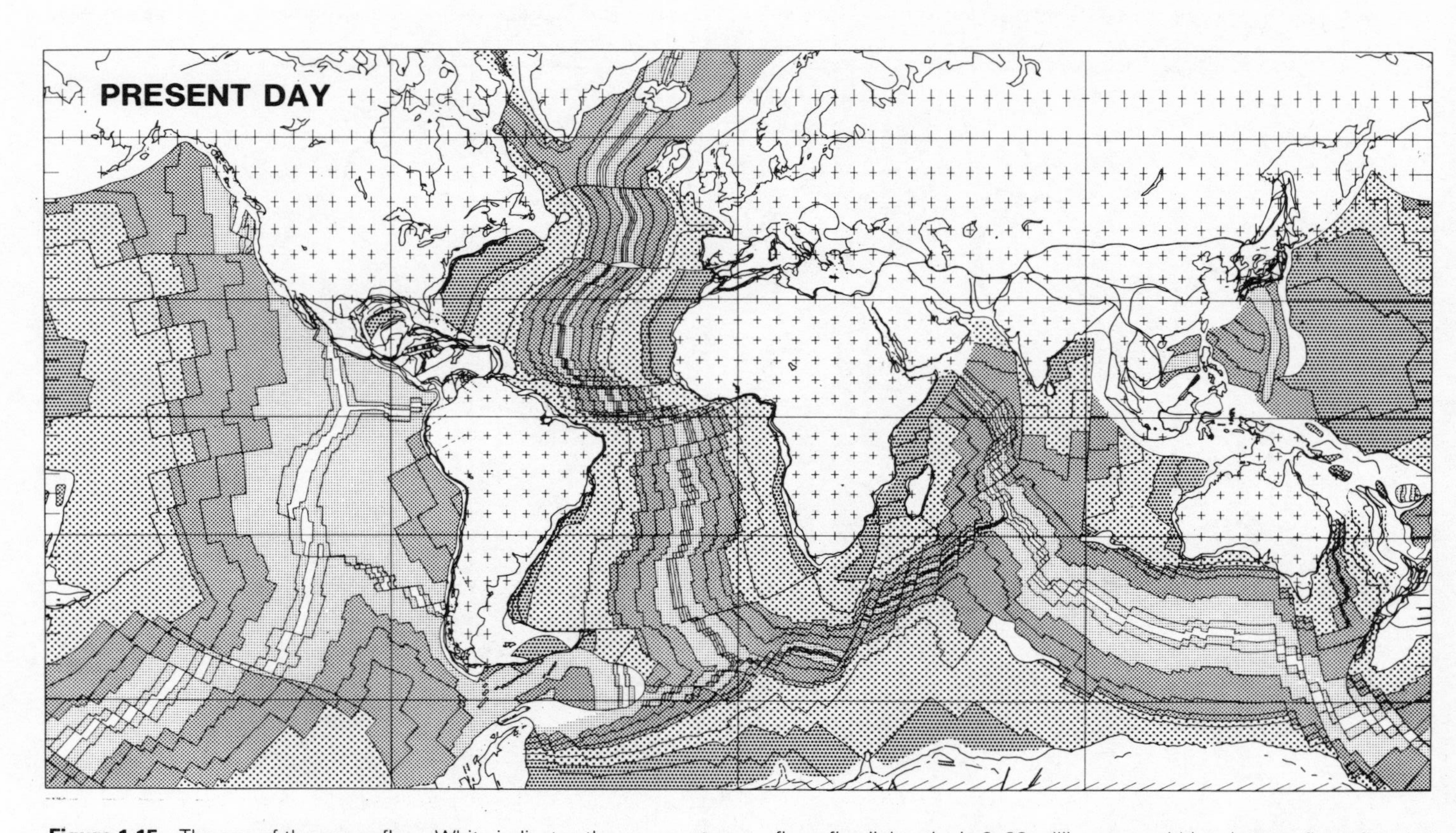

Figure 1.15 The age of the ocean floor. White indicates the youngest ocean floor, fine light stipple 2–23 million years old (myr) ocean floor; fine dark stipple 23–59 myr ocean floor; coarse light stipple 59–84 myr; coarse dark stipple 84–144 myr ocean floor (from Scotese *et al.* 1988. *Tectonophysics* **155**, 27–48).

difference is even greater: one half of the planet is, in effect, continent (but that is another story). In contrast, Venus seems not to have clearly defined continents, or at least not continents in the sense that we have continents on Earth.

THE IMPORTANCE OF WATER

The difference between Earth and Venus may partly be explained by the presence of water on Earth. The hydrothermal systems at mid-ocean ridges add water to the new oceanic crust. This water plays a very important role in the melting processes that produce the volcanoes at subduction zones. The rafts, which constitute the continents, were all formed or have been reworked above subduction zones. Without water, subduction zones would be very different and might not exist in their present form. If so, the continents could not have been constructed, at least in the way they exist on Earth. Without water, there might have been no continents.

Water also helps to destroy the continents by eroding them. Since the continents effectively float on the viscous interior of the Earth, in a process analogous to the floating of blocks of wood in a bucket of water (a process called **isostasy**), the base level of erosion (which is sea level) is one of the controls on the thickness of the continents. This means that there is a relationship between the depth of the oceans and the thickness of the continents. Deep oceans imply thick continents. On the modern Earth the oceans are about 3.5 km (2 miles) deep on average, and in consequence, because of the relationship between erosion and water depth, the continental granitic rafts are eroded down so that they have an average crustal thickness of about 35 km or 20 miles.

We can now begin to appreciate the critical role of water in the geological make-up of the Earth's face. There are really four water cycles, nested into each other. The first is the rain cycle. Enormous volumes of water are evaporated off the face of the ocean, fall as rain and run back to sea, meanwhile eroding the continents and transporting sediment out to the coast. On a longer timescale, in the second cycle, water is introduced into the oceanic crust by the hydrothermal systems at the mid-ocean ridges. Most of the water immediately returns to the sea, but a small amount leaves the cycle and remains in the oceanic crust, which gains water, to become hydrated. In the third cycle this water of hydration is eventually carried down the subduction zones, perhaps 60 million years later, together with that part of the sediment eroded off the continents which was dumped onto oceanic crust. Much of the water and sediment that was carried down is then returned to the surface, over a period of a few million years, by way of volcanism and the intrusion of granites and similar rocks. A small amount of water may be carried down the subduction zones to the deep interior.

Finally, there is also a slower process involving water from deep in the

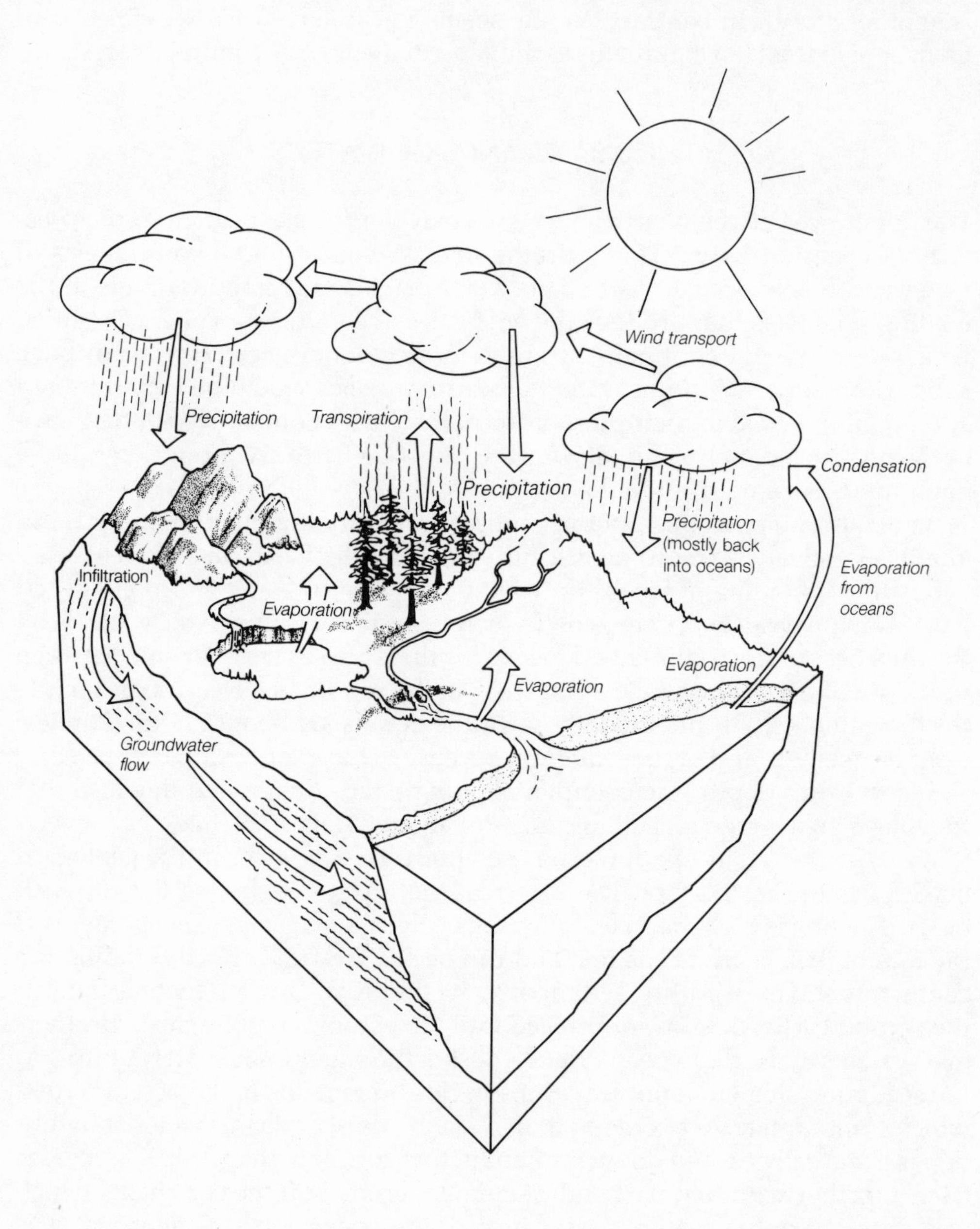

Figure 1.16 The hydrological cycle (courtesy of Department of Geology, University of Saskatchewan).

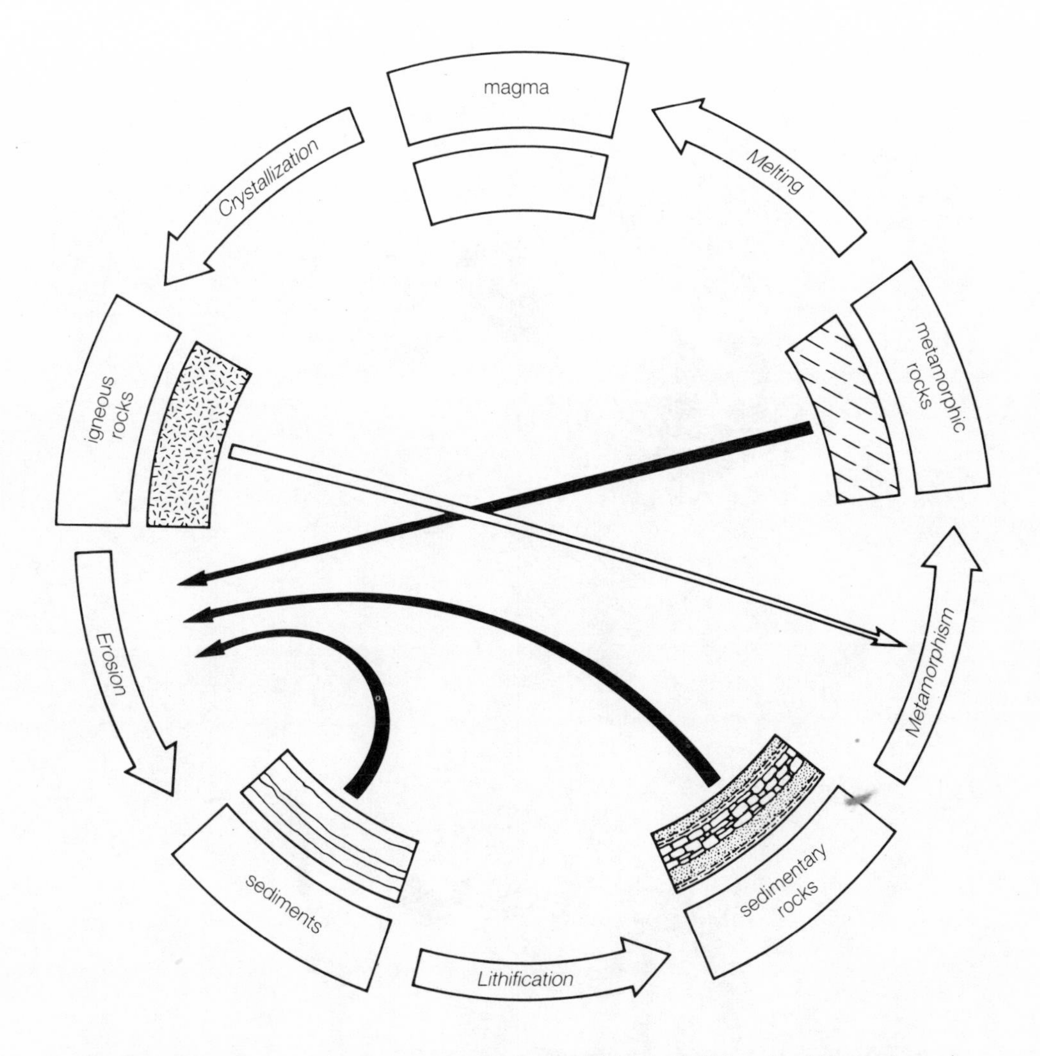

Figure 1.17 The rock cycle (courtesy of Department of Geology, University of Saskatchewan).

Earth's mantle. This water is erupted, or degassed, at the mid-ocean ridges and also at deep-rooted volcanic centres within plates such as Hawaii. The degassed water enters the oceans and is added to the external system. One estimate is that more water is now carried down into the interior by subduction zones than is degassed, so that there is, today, a slow net loss of water from the surface in this process. Perhaps about one part in two billion of the total ocean is lost each year. A small amount of water is also exchanged with outer space: some is lost to space from the top of the atmosphere in the same way that Venus lost its oceans, and some is gained from small comet-like bodies hitting the Earth. The net effect of all this cycling is very controversial: a small difference between two poorly known large numbers. It may be that the oceans were

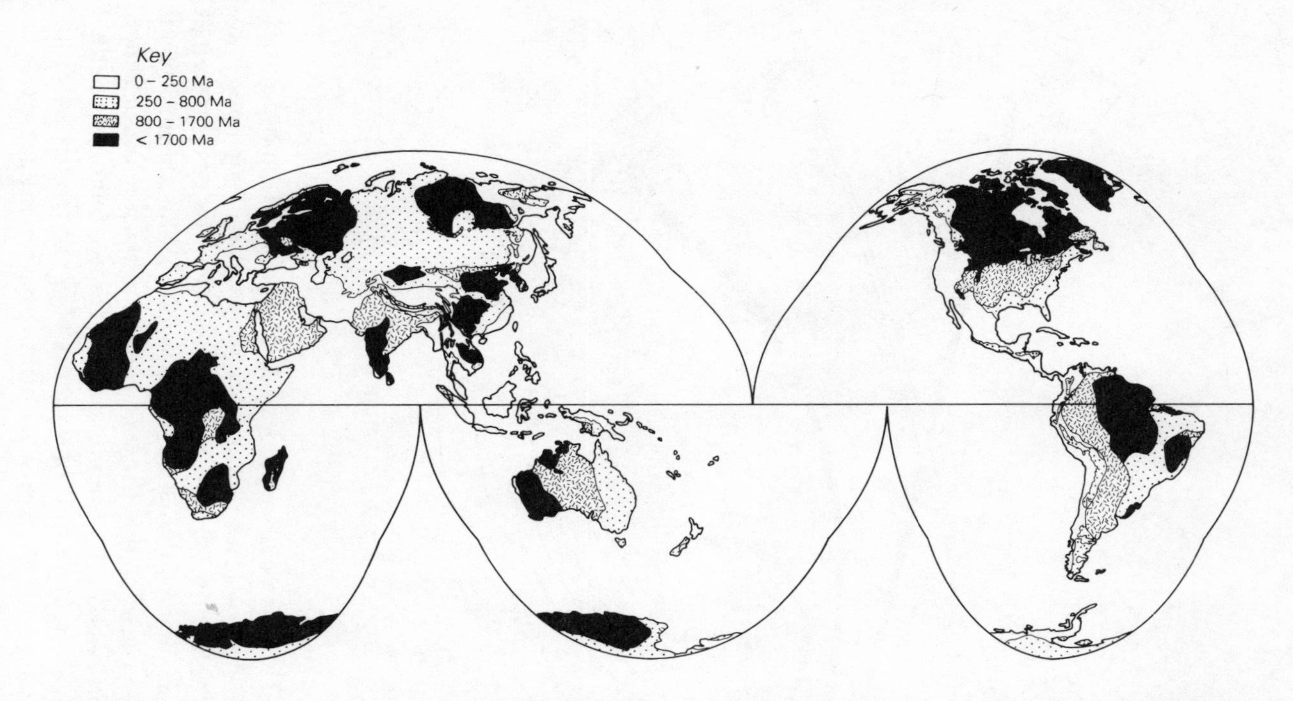

Figure 1.18 The age of the continental crust (modified after Sclater *et al.* 1981. *Journal of Geophysical Research* **86**, 11535–52 by Fowler, C.M.R. 1990. *The solid Earth*. Cambridge: Cambridge University Press).

twice as deep in the distant geological past, and the continents thicker. Perhaps the oceans may finally be lost billions of years from now.

One of the central features in all this is the hydrothermal circulation at the mid-ocean ridge. Venus is air-cooled, like an old Volkswagen Beetle, while Earth is a water-cooled Rolls-Royce. The hydrothermal circulation is the radiator of the Earth's engine. It rapidly cools oceanic crust; it causes oceanic crust to be hydrated; it helps subduction to operate and thus helps the continents to exist; and it is an important control of the chemical composition of seawater.

FURTHER READING

Broecker, W. 1985. *How to build a habitable planet*. New York: Lamont Doherty Geological Observatory of Columbia University, Eldigio Press.
Brown, G.C. & A.E. Mussett 1981. *The inaccessible Earth*. London: Allen & Unwin.
Cloud, P. 1978. *Cosmos, Earth and man*. New Haven: Yale University Press.
Fowler, C.M.R. 1990. *The solid Earth*. New York: Cambridge University Press.

2
The beginning of life

WHAT IS LIFE?

The first of all geological puzzles is how life began. We have no real idea of the answer. Any living organism, even the simplest, is an enormously complex array of chemicals. To make life accidentally would need an extraordinary and unlikely series of events. The biologist, Francis Crick, in his book *Life itself*, pointed out that the chance of life beginning on Earth is so remote that one must demand a miracle, whether it be the deliberate seeding of the planet by some distant life form (Crick's preference), or the more conventional appeal to an extraordinarily improbable event. Theologically and rationally there is no reason against the latter explanation being true: life may well be the result of an extremely unlikely event. Was there divine intervention? The psychologist, Carl Jung believed in the Latin saying: 'Called or not called, God will be present'. Other scientists feel that natural processes will inevitably produce life on a habitable planet. The discussion remains metaphysical, not in the domain of natural science.

What we can do, though, is assume that once, on a rather improbable planet (Earth), a wildly improbable event (the beginning of life) took place. Having accepted that as an axiom, because we are here, we can speculate about the special events that created life.

Normally, in investigating a phenomenon one must first define it. Yet neither science nor philosophy can define life: it eludes our attempts to pin it down. What is life? It is not reproduction, because crystals reproduce themselves after a fashion, while eunuchs do not. It is not some funny property of carbon atoms, as we can, in principle at least, conceive of some electronic analogue of a very simple bacterium. We cannot even draw the line between life and non-life – is a virus life, or is it a naked string of chemicals? At some times viruses seem to be living; at other times they can be stopped, inactive. We can crystallize the genetic material, DNA, just as we crystallize salt, yet DNA is the key of life.

Many people have tried, unsuccessfully, to define life. The political philosopher Engels, a friend of Marx, considered it to be the mode of action of albuminous substances, a definition which admits under the title 'life' such oddities as breaking eggs and making omelettes. The famous English scientist, J.D. Bernal, who otherwise thought clearly, came up with a similarly useless definition, that life is the self-realization of atomic electron states, a definition aimed more at denying God than at explaining life. Thermodynamics helps us to come closer: all life shares the property of increasing the local order by making the environment around it more chaotic. Life is growth. It can exist only in time, creating something that is locally special by exploiting the environment. In effect, it mines the future to enrich the present. Perhaps Cardinal Newman, a theologian, was closest to it, although he thought in theological, not thermodynamic terms. He lived by the maxim: 'Growth, the only evidence of life'. Life is sustained, growing unbalance, or 'disequilibrium'. Even in old age, our cells are still working, processing, sometimes growing. Chemical balance, or equilibrium, means death. Indeed, in the progression from the simple bacterium living in a hot pool to a man walking on the Moon, the degree of sophistication of a living organism can be measured by its degree of disequilibrium from a natural environment. Life, like a Western economy, cannot stay still. It must grow. If it stops, it dies and attains equilibrium as a corpse.

Yet there is still something elusive. How do we distinguish between growing crystals and life? We still cannot define life. At the centre of the problem the mystery remains.

THE MAKING OF LIFE

Life creates an ever-growing degree of local order. It does this by self-replication. The mathematician von Neumann showed that any self-replicating machine must have both a *list* of instructions to direct replication and a set of *organs* to read the list and use it to reproduce. In modern life, the list of instructions is made from either DNA (deoxyribonucleic acid) or RNA (ribonucleic acid), which are organic compounds made of carbon (C), oxygen (O), nitrogen (N), hydrogen (H) and phosphorus (P). Most life, except the very simplest, keeps the list in DNA and then transcribes the instructions into lengths of RNA, which are used as the working copy for making the organs. Nucleic acids are made from chemical building blocks known as sugars, bases and phosphates. The **sugar** in RNA is a compound called ribose ($C_5H_{10}O_5$) and in DNA it is deoxyribose ($C_5H_{10}O_4$). The **bases** are nitrogen-containing compounds, with ring-like shapes, known as purines and pyrimidines; their names are adenine, guanine, cytosine and thymine in DNA, and adenine, guanine, cytosine and uracil in RNA. **Phosphate** is made from a phosphorus atom surrounded by oxygen. The combination of sugar, base and phosphate is known as a **nucleotide**. The acids, which are strings of sugar alternating with phosphate, with a

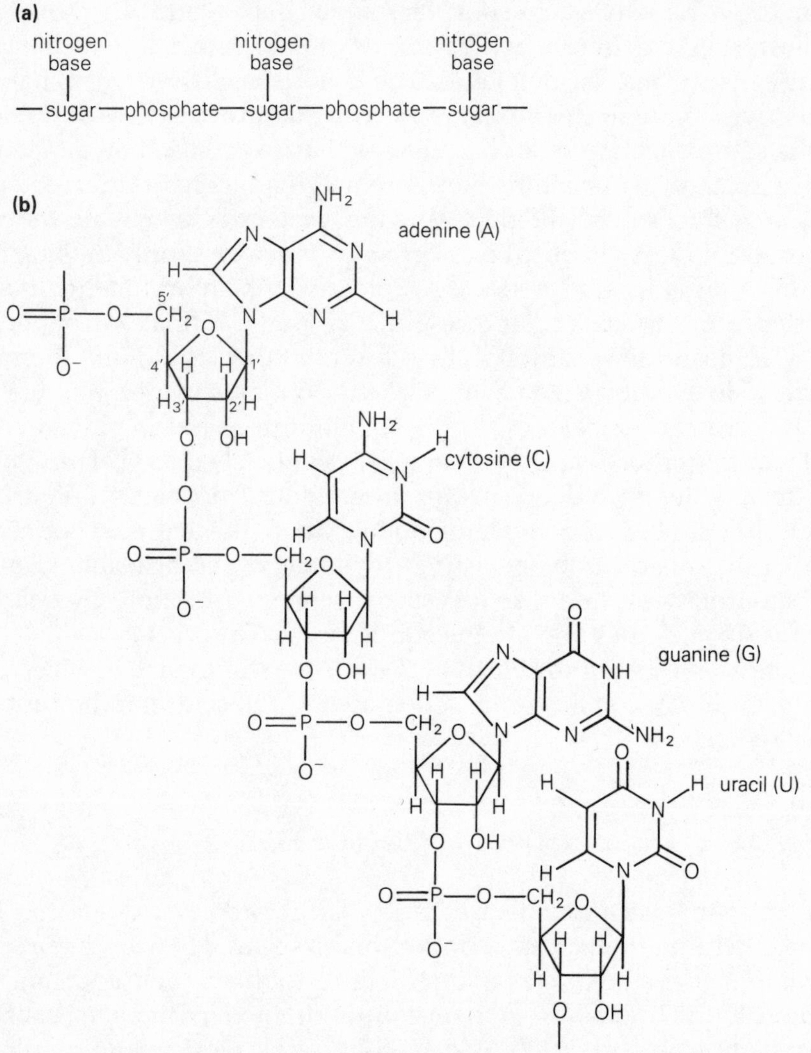

Figure 2.1 The structure of ribonucleic acid, RNA. (a) How the molecule is put together. (b) An example of a length of RNA, illustrating the four nitrogen-containing bases: adenine (A); cytosine (C); guanine (G); uracil (U). The position of carbon atoms is shown by 1′, 2′, 3′, 4′, 5′ in the top sugar ring.

base attached to each sugar, are **polynucleotides**. These are the fundamental information stores of all living things. RNA is a string of nucleotides, DNA is a double string wrapped together as a spiral, a double helix.

In contrast, the organs of life are mostly **proteins**. They do the work of maintaining the conditions around the list so that the organism can replicate, and also help in copying the list itself. Proteins are chains of **amino acids**. These are compound acids containing a carboxylic group, COOH, and an amine group, NH_2. There are 20 common amino acids in the proteins of living organisms. In

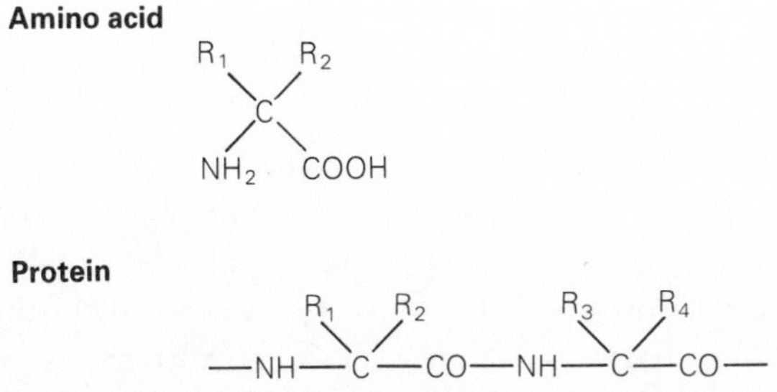

Amino acid

R_1 R_2
C
NH_2 COOH

Protein

R_1 R_2 R_3 R_4
—NH—C—CO—NH—C—CO—

Figure 2.2 Amino acids and proteins. The top diagram shows the typical structure of an amino acid. The lower diagram shows a length of protein, made of amino acids joined together. R represents different side chains. In life, each carbon atom at the centre of each amino acid is connected to a hydrogen atom (R_1, R_3) and one of 20 different side chains (R_2, R_4).

making a protein, amino acids are linked together in CO–NH bonds, known as peptide bonds, to form a long polypeptide.

The **cell** is the basic building block of life. Many organisms, such as bacteria, have only one cell. Others, such as humans, have many cells. Proteins are the workhorses of the living cell. They are the organs of the cell. Some proteins help in the replication of the nucleic acid list, others act in the construction of the cell. The cell also has a bag or membrane around it, built of small molecules (phospholipids) made of fatty acids, glycerol and phosphate. The membrane protects the local order within the cell from the surrounding environment and enables the cell to create special conditions on the inside. Both the list of instructions and the organs are confined within the bag and do not drift away from each other.

THE CHICKEN OR THE EGG?

In modern cells, the list of instructions is written in the DNA sequence. Lengths of RNA are then made by copying from the DNA. These lengths of RNA serve as the working instructions – almost as if an architect had a massive master plan of a building in his drawing office but used a collection of blueprints on the building site, one blueprint for each room. In this way the original master plan could be kept clean, while the blueprints could be used on the construction site by the builders until the plans became so dirty and torn that a new copy was needed. But the analogy should not be pushed too far; DNA and RNA are much more sophisticated than simple blueprints.

The RNA lengths are then read and proteins are constructed out of the amino acids specified by the RNA instruction list. Some of these proteins act as biological helpers that aid the process of copying and replicating the instruction list: these biological helpers, or **enzymes**, are molecules of very specific shape that offer surfaces on which the nucleic acid list can be duplicated. Thus we have one of the paradoxes of life: in order to make a protein, a code of instructions from a nucleic acid is needed, but to replicate the nucleic acid the aid of a protein is required. This is exactly the old problem: which came first – the protein chicken, or the nucleic acid egg?

Most of those who speculate about the origin of life recoil from the improbability of chicken and egg being inorganically created simultaneously. As a result, many models of the origin of life invoke one or the other: protein first or nucleic acid first. These models attempt to explain how one component (either protein or nucleic acid) was formed and began life, then try to explain how the other component evolved. Some models are different, and suggest that neither the chicken (protein) nor the egg (nucleic acid) was first, but that some primeval progenitor came first, which was later displaced when more efficient organisms evolved that were based on both nucleic acid and proteins.

Finally, there are the models that invoke the idea of panspermia. This is the idea that life appeared once, somewhere else in space, and then spread itself around the universe. The advantage of this idea is that it allows more time for life to develop – billions of years, not the few hundred million available on the early Earth; the disadvantage is that it simply transfers the difficulty elsewhere and then further demands a spreading agent. The idea of panspermia we can either take or leave, but it is hard to debate the likelihood that it is true. Crick's book, *Life itself*, is an excellent exposition of the panspermia hypothesis. Let us, instead, assume that life *did* begin on Earth (although we should never forget that this assumption may not be true) and examine the possibilities.

The soup?

Many people imagine that life originated in some sort of primeval soup, an ocean teeming with assorted organic chemicals that happened to fall in from space on board meteorites, or which were somehow synthesized on the planet. Above this soup floated an atmosphere of methane, ammonia and the like, pierced occasionally by lightning flashes. In this soup, droplets of organic chemicals formed and somehow evolved into life.

Unfortunately, much of this picture seems too simplistic to the geologist. First, did the soup exist? We know a little about the chemical processing that must have taken place in the oceans, even billions of years ago. The hydrothermal systems at mid-ocean ridges have already been described. On the young Earth (say over four billion years ago) the cycling time, that is the time taken for a volume of water equal to the total oceanic volume to pass through the hydrothermal system, would perhaps have been of the order of one million years, at a rough guess. Geologically speaking, this is a short time. During hydrothermal circulation, the seawater and all it contains is heated to high temperatures (sometimes 650°C or more, though usually less), at pressure. These conditions are enough to break up dissolved complex organic chemicals into much simpler molecules. Any primeval soup would have gone through this process, and so it is unlikely that early seawater had high concentrations of dissolved organic chemicals that had landed on the Earth on board meteorites and slowly accumulated. Perhaps the chemicals floated on top of the ocean and escaped destruction? If so, they would have to escape the perils of the Sun's ultraviolet irradiation. The primeval soup must have been very thin indeed: could the interesting molecules have found each other in such dilution, in order to come together to make life?

The hypothetical atmosphere of methane and ammonia is another construct that is vulnerable to solar irradiation, in the presence of carbon dioxide, nitrogen and water. Methane in the modern atmosphere is mostly biologically made. In the modern atmosphere it is destroyed within roughly eight years by hydroxyl, OH, ultimately derived from water. The Earth's early atmosphere would have been different with different chemical reactions, but in it, too, any volcanic or primeval methane would have been destroyed rapidly. Only on the outer planets are methane-rich atmospheres stable. Most probably the early Earth had a chemically mild atmosphere dominated by carbon dioxide, water, and probably nitrogen, perhaps with some oxygen left over from the destruction of water by solar radiation at the top of the atmosphere. This air lay above an ocean that was rich in chemicals derived from hydrothermal systems but probably poor in organic chemicals, not a soup. But if life did not begin in a soup, where did it start? The next most improbable place to look is the surface, to the rocks and clays on the seabed.

The crystal pattern

J.G. Ballard once wrote a work of science fiction in which all life turns to crystal: there is something in common between the fabric of life and the growth of a crystal. Both life and crystals are information-bearers, replicating a pattern. The British chemist A.G. Cairns-Smith has suggested that life developed from crystals, being originally based on the replication of clay crystals. In the Cairns-Smith model, life began through the influence of natural selection on the growth of inorganic crystals. If so, the first **genes**, or carriers of genetic information, were clay structures, replication taking place by the accidental detachment of layers in the clay lattice, which served as nuclei for the growth of new daughter molecules. Later, organic chemicals were incorporated into the structure of the replicating crystallites, and competition favoured those systems that were more adaptable because they used organic molecules to carry out their functions. Nucleic acids (RNA and DNA) then evolved, taking over as the basic information list making up the genes of the organism and, under the pressure of natural selection, the original clay-mineral component was dispensed with entirely.

This is a fascinating and powerful hypothesis. It is as likely to be true as any other idea and is well worth investigating. But there is always the problem of how a daughter organism could be detached from a crystallite, except by irreproducible accident. One of the oldest principles in Western science is Occam's Razor: 'hypotheses should not be multiplied unnecessarily', *essentia non sunt multiplicanda praeter necessitatem*. The razor is not infallible, and it sometimes cuts off truth, but it frightens us away from ideas that demand too many steps. It is worth investigating the organic-only hypotheses.

The chicken: proteins

Perhaps the most popular hypothesis is the notion that life began with proteins. In this model, a self-replicating organism made of proteins was accidentally put together, and only later did the nucleic acids evolve to carry the list of instructions. To return to von Neumann's self-replicating machine, in the protein-first model the organs and the list consisted at first of a series of amino acids; later, during the accidents of evolution, the properties of nucleic acids were discovered by life, leading to the incorporation of nucleic acids (RNA and DNA) as the information store because of their greater usefulness or, more correctly, fitness in natural selection (see Chapter 4).

This hypothesis is attractive because it is rather easier to imagine the inorganic synthesis of proteins than it is to imagine the inorganic and accidental synthesis of nucleic acids. It has been suggested that in some sort of special local environment, especially with clay minerals or other crystals to help as catalysts, proteins could have been constructed from simple chemicals. For example, in the complex hydrothermal chemical factory near a volcano, amino acids might have been made and then linked in primitive protein-like molecules, in the same general process that was breaking up any planetary organic soup.

The flaw in the protein-first hypothesis, however, is the difficulty of replication. Proteins do not replicate themselves in a way that carries useful information. The attraction of the clay-first hypothesis is that replication is initially straightforward, though the amount of information transferred is small: later, more complex replication evolves. In contrast, in the protein-first hypothesis, complicated information has to be transferred from the beginning, and there is no obvious way to assemble a new protein without a pre-existing chemical factory.

The egg: RNA

Finally in the search for the origin of life, we turn to the egg. In modern organisms, the two types of nucleic acid involved in carrying the list of instructions are DNA and RNA. The central store is DNA, the double or two-stranded helix. This is a secure record, with two complementary copies – one on each strand of the helix – so that accidental errors on one strand can be corrected by reference to the other. If ultraviolet light causes sunburn and damages one strand of the DNA in our skin cells, the cell can often repair the list by copying back from the other strand (sometimes, unfortunately, the DNA is so damaged and changed that it creates cancer cells). DNA is unwound and mixed in reproduction, including sexual reproduction: 'important biological objects come in pairs'.

But for all its marvels, DNA may not be the original nucleic acid. Many scientists think that RNA (ribonucleic acid) came before DNA. RNA is single-stranded and it carries the information from the DNA to the ribosome, which is the reading head of the genetic tape recorder. *Enzymes,* which are helper molecules generally made of protein, enable the RNA to copy and transfer stretches of information. The process needs the list of information and also the help of the protein enzymes, themselves made from the instructions in the list. Nucleic acids today direct the construction of proteins some of which then help in replicating the nucleic acid, but the first replicating system could not have been so complex, unless the whole factory was invented at once in a place with a large supply of all the right chemicals. Almost certainly, if life began with RNA, the molecule must have replicated *without* the help of protein enzymes. An egg (nucleic acid) must have produced another egg without the help of a chicken (protein).

Until a few years ago, this objection seemed as fatal to the nucleic acid-first hypothesis as the replication problem is to the protein-first hypothesis. Nucleic acids need proteins to help catalyse their replication.

Chemical reactions in the cell are promoted by the presence of helper enzymes, or catalysts. These catalysts provide specific sites at which the chemical reactants can meet. The catalyst plays no intrinsic part in the reaction. It simply helps to bring the participating reactants together. Without the catalyst, however, the reaction would probably never have taken place. No inorganic catalyst can match the exquisite exactness and specificity of organic enzymes, and it seemed until recently that all enzymes were made of protein. However, it has been found that under certain special circumstances, RNA can detach lengths of itself to act as enzymes. This means that it is possible that RNA may not only contain a list of instructions, but it may also be capable of detaching an organ to help in the replication of the list.

This recent discovery, of RNA-enzymes, or *ribozymes,* is at the centre of the notion that RNA made the first living organisms. If the right molecule of RNA could somehow be constructed, it could in certain circumstances splice out a portion of itself and this portion could act as a catalyst, or enzyme. The right circumstances appear to include a mildly alkaline environment, a temperature near that of our own blood (a little below 40°C, 100°F), and the presence of various cations. In the correct conditions, under attack by guanosine (an organic compound that is a nucleotide based on the nitrogen base guanine), a section of RNA is cut out of the parent molecule and the detached section can have the properties of an enzyme; it can behave as a catalyst that, in principle at least, can aid in the replication of the molecule.

In modern life the organs of the cell are protein-based, but this strange property of RNA allows us to imagine an RNA-world, in which both list and organs were made of RNA and initially protein played no role in replication. The RNA is a list of instructions, but it can also act just as proteins behave in most modern replication reactions, as an enzyme or catalyst. Indeed, the ability of RNA to detach a length of itself as a daughter molecule of the same substance is an ability that is the essence of life. RNA is a subtle and beautiful molecule, not perhaps as spectacular as the DNA double helix, but with properties that may lie at the very core of the enigma of life.

Not only can RNA detach lengths of itself, it can also evolve. If RNA molecules existed in a micro-environment, information could be transferred from one molecule to another by detaching lengths of RNA from one molecule under certain conditions and then, under slightly different acidity, inserting them into another nearby RNA molecule. By this process infectious transmis-

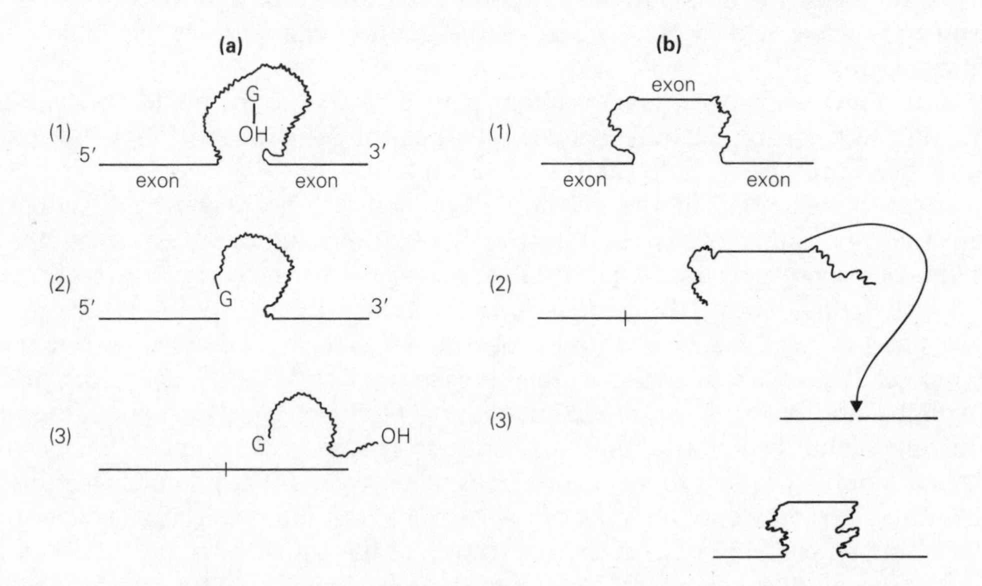

Figure 2.3 The unusual abilities of RNA. Diagram (a) shows how an RNA molecule can detach an intervening sequence. Initially, in this example, the RNA contains two sequences, called **exons**, separated by the sequence to be detached. Under attack by guanosine the intervening sequence is cut out and the two exons are spliced together. Numbers 5′ and 3′ refer to the two ends of the RNA sequence, defining carbon atoms in the ribose sugar to which the phosphates are attached. Diagram (b) shows how information could be exchanged in an RNA world. In (1) an RNA molecule has two sequences, or **introns**, with properties like those of the intervening sequence in diagram (a). These introns, and the exon between are detached. The remaining exons in the parent molecule are then spliced together, and the detached introns and exon cleave another RNA molecule (see arrow) and are spliced into it, thereby transmitting information from one RNA molecule to another.

sions of genetic information could occur from one RNA molecule to another, in a molecular analogue of sexual reproduction.

RNA is thus an extremely powerful molecule. It stores genetic information. It is able to act as an enzyme and to help in replication. Any process involving the exchange of information runs the risk of a mistake: pieces of data can be added, deleted, or copied wrongly. Most mistakes impair the usefulness of the data, but sometimes by accident the mistake improves the information. These improved versions are more likely to survive, as they are better fitted to the environment. This favouring of the better suited is natural selection. Once made, a self-replicating RNA molecule would be subject to natural selection: it would evolve. The flaw in the RNA-first hypothesis is that it is very difficult to imagine how, on Earth, an RNA molecule could be constructed by accident out of the chemical components available in the environment. RNA, even in short lengths, is a very complex molecule, and it is difficult to construct its components.

It is possible, despite the difficulty of constructing it, that an RNA molecule was the universal ancestor. To summarize, we really do not know. Life could have begun with RNA, but the protein-first model is also possible, as is the clay-mineral model of Cairns-Smith. There is also always the off chance that the planet was seeded from space. But, to move on, on the assumption that it did happen on Earth, where did it happen?

THE BIRTHPLACE OF LIFE: DARWIN'S 'IF'

In 1875, a group of scientists on board the research ship HMS *Challenger* observed a hydrothermal system on the slopes of a volcano in the Philippines. They recorded the life in the hot springs of the system and speculated whether 'green algae of some considerable complexity may have commenced life on earth in its early history' growing in water 'which may have been strongly impregnated with various volcanic gases and salts'. Could this have been the environment in which life was born? This speculation was shared by Charles Darwin who was in touch with Moseley, one of the *Challenger* scientists. He wrote, in a letter to another colleague, Hooker:

> 'But if (and oh, what a big if) we could conceive in some warm little pond, with all sorts of ammonia and phosphoric salts, light, heat, electricity, etc., present that a protein compound was chemically formed, ready to undergo still more complex changes . . .' Could life be made in some warm little pond? As Darwin noted, we shouldn't expect to see it happen today for 'at the present day such matter would be instantly devoured or absorbed, which would not have been the case before living creatures were formed'.

Since the days of the *Challenger* expedition, there have been many other ideas. The most popular idea is probably the notion that life began in the sea, the organic primeval soup. The grave objections to this hypothesis have been discussed above. But there is the further objection that even if a molecule capable of replicating did form, it would not have been able, by chance in the sea, to meet chemicals from which it would make new daughter molecules. It would have been broken up before it bumped into the right collection of components. Even more critical, a list molecule needs organs to help it replicate. How could a molecule in the soup keep its associated organs, or catalytic helper molecules, close by? They would surely drift away and become useless, eventually to be destroyed by ultraviolet light or hydrothermal cooking.

To find life's birthplace, perhaps we must instead turn to a local environment, of the sort that Darwin suggested in the 'warm little pond'. Innumerable possibilities have been proposed. Many scientists feel that because life itself is a wholly improbable phenomenon, an extraordinary environment could have been the scene of its birth. Examples of such unlikely but plausible settings include a volcano under an icecap (there are modern examples in Iceland), or an aerosol droplet in the atmosphere, or a small cold pond hit by a meteorite that was rich in organic matter.

We can attempt to constrain the speculation by setting up criteria that the postulated birth-setting of life must satisfy. First of all, there must be an adequate supply of the small chemical building blocks out of which life is made. Secondly, an aid to putting them together would be helpful – some sort of catalyst. Finally, some process of physical or chemical changes must have been taking place to provide a thermodynamic driving force to make the assembly reactions go forwards.

What were the small chemical building blocks? To recapitulate, nucleic acids are made of nitrogen-containing ring compounds, sugar and phosphate. Proteins, which are chains of amino acids, include carboxylic groups (COOH) and amino groups (NH_2) linked in a chain with associated carbon-containing groups. Innumerable experiments have been carried out in an attempt to discover how these chains could have been put together: the mystery remains, but chemicals such as methane (CH_4), ammonia (NH_3), ribose sugar and phosphate were probably involved, as well as cyanide (HCN) and formaldehyde (H_2CO).

Here lies a great problem: as discussed above, methane and ammonia, in particular, may not have existed in any significant quantity in the ocean or atmosphere. However, methane and ammonia occur in hydrothermal systems in sub-aerially exposed sections of mid-ocean ridges as in Iceland, or around volcanoes on continents. Most modern methane is generated in biologically controlled settings, such as muds or the digestive system, but around volcanoes and in geothermal systems carbon and nitrogen compounds can also be reduced to methane and ammonia. These carbon compounds may or may

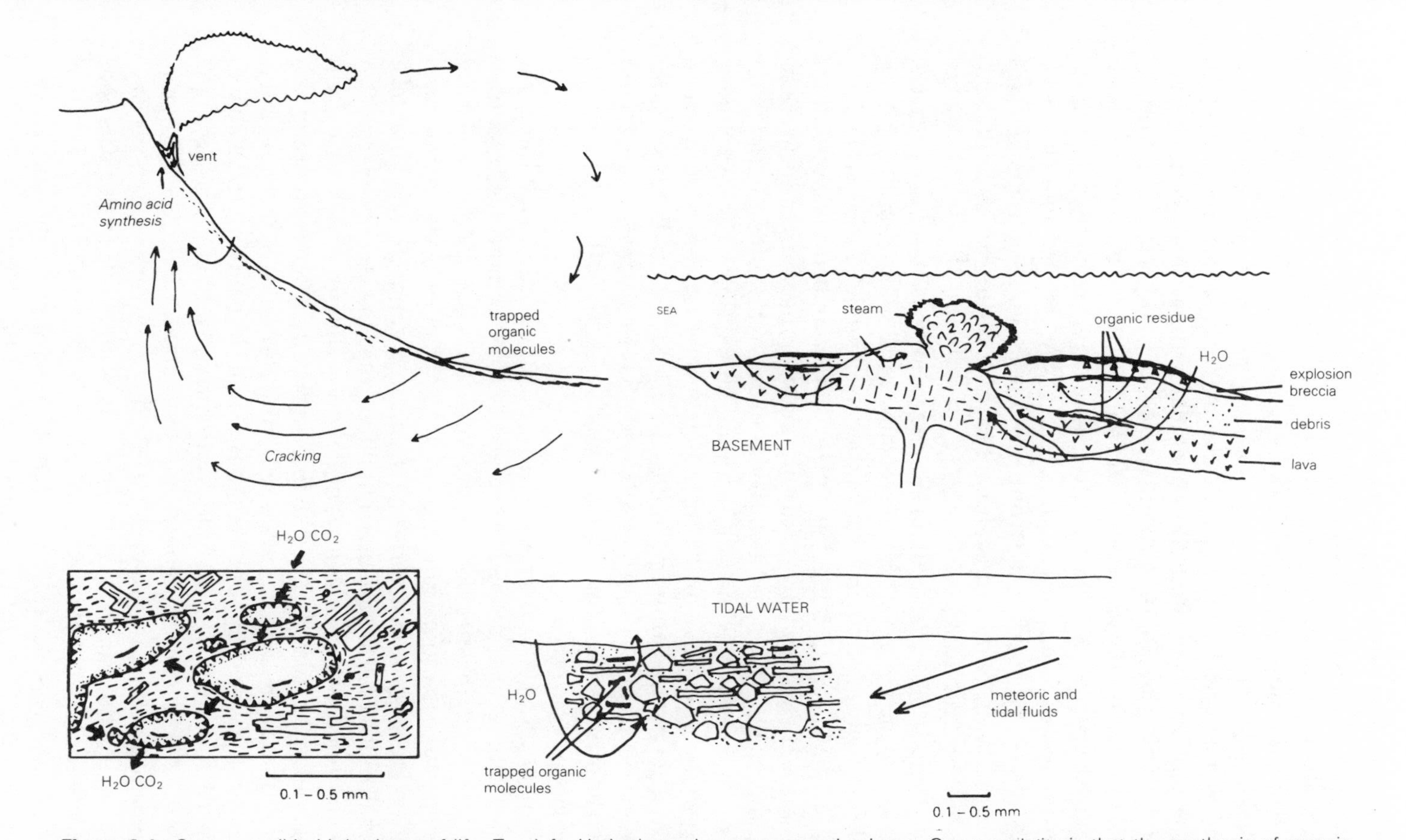

Figure 2.4 Some possible birth places of life. Top left: Hydrothermal system around volcano. One possibility is that the synthesis of organic molecules took place around mid-ocean ridges. This is unlikely, however, as the conditions would have been very hot and very acid. Volcanoes in shallow water, or exposed subaerial volcanoes may, however, have been sites of synthesis. Top right: Systems around shallow-water explosive acid volcanoes. Seawater enters the rock, which is made of porous volcanic debris; as the water nears the hot volcanic centre it warms, becomes less dense and rises, to exit from hot vents or as steam explosions. Bottom left: Movement of H_2O, CO_2 and other gases such as methane or ammonia through lava, with cavities lined with crystals of alteration minerals such as clays, tubular iron oxide compounds and zeolites, where molecules could be synthesized from the mix of components available as temperature and acidity varied. Bottom right: Movement of fluid in shallow tidal settings.

not be of organic origin; some may be derived from gases from the Earth's interior, others may be introduced from the surface. On the young Earth, in the Hadean Eon, meteorites may even have supplied methane.

Sugars are also important. It is possible that around volcanoes, for instance in gas-rich vents known as fumaroles, sugars were synthesized via simple compounds such as formaldehyde. Perhaps the nitrogen-containing bases of nucleic acids could have been formed in a similar setting, but it is difficult to imagine a geologically plausible environment. Most reactions required to make the bases would have involved hydrogen cyanide (HCN), but cyanide is an unstable and thus very transient chemical in natural environments. Hot natural reactions can make cyanide, but it is an unusual and improbable chemical. Indeed, the absence of cyanide is a strong argument against the nucleic acid first hypothesis: without cyanide it is difficult to imagine how RNA could have been synthesized inorganically.

Catalysts, or helper molecules, are also needed to help the first synthesis: whatever the setting, some at least of the reacting molecules would have been rare chemicals, unlikely to bump into each other accidentally. The best catalysts are those made by life – what could have played this role before they existed? The answer may be that minerals provided the sites for the first synthesis. Minerals that could perhaps have acted as catalysts include clay minerals and other minerals called zeolites as well as iron oxide minerals. Some clays may have been able to bind nucleotides and promote the construction of long-chain molecules. Clays today are common. In the earliest part of the Earth's history they would have been rarer, though they would have occurred in weathered volcanic rocks, around most of the hydrothermal systems, both on land and in submarine ridges.

Another mineral group, known as zeolites, includes impressive inorganic catalysts, which can be so selective in their ability to bind molecules that they come close to the exquisite specificity of proteins. They also possess a second fascinating property in that they can act as molecular sieves. Some zeolite minerals themselves are shaped as a series of large cages connected by channels. They allow small molecules to pass through while trapping larger molecules. A small rock cavity lined with zeolites would allow a flow of water, or carbon dioxide, methane, ammonia and the like to pass through, while trapping any larger organic molecule. However, only a few natural zeolites (which are, interestingly, found in hydrothermal settings) have cages big enough to be able to host nucleic acids (larger than simple inorganic compounds) or proteins involved in life. Furthermore, zeolites are not good catalysts of organic reactions in which much liquid water is involved. They work well in the dry, but not in the wet, like the paper underwear briefly fashionable some years ago. A third candidate for the role of mineral catalyst is a family of minerals known as tubular iron oxide hydroxides. These too have very interesting structures that may have provided sites for the synthesis of the first life.

A different clue to the location of the first synthesis may be given by the conditions under which RNA can detach a helper molecule, or ribozyme, from itself. If the notion that RNA was the parent molecule is correct, then the conditions in which daughter lengths are detached – temperatures around those of a warm bath, mildly alkaline water, and with magnesium ions present – might be those of the first RNA world. This may be our most powerful guide to discovering the birthplace of life, but *only* if the RNA-first idea is right. All this is guesswork, not evidence. We have no real knowledge.

THE HYDROTHERMAL POSSIBILITY

The imaginings of the men of HMS *Challenger* and Darwin's 'warm little pond' have ever since haunted geological (though not chemical) thinking about the birthplace of life. More recently, geophysicists and geologists in the early 1970s predicted that vast hydrothermal systems existed on the mid-ocean ridges. These were subsequently discovered, complete with complex living communities. The discovery has made the ghost more real. Many of the bacteria living around hydrothermal systems are **archaebacteria**, a type of bacteria that is markedly different from most of the more common bacteria on the Earth's surface. The mid-ocean ridge community does not, at first impression, seem to need the Sun for its energy source: the organisms live directly off the chemical differences between the hot emerging hydrothermal fluids and the surrounding environment. Could this mid-ocean ridge community be some primeval relict of the birthplace of life?

The answer, sadly, is probably no, except in the broader sense that all living communities are descendants of the first ecosystem. Just as the Kalahari Bushmen are not primitive, but have evolved a complex and successful culture able to cope with their extreme environment, so probably (we are not certain) have the deep hydrothermal communities developed over the aeons. They have evolved differently from surface life, but nevertheless in a complex and successful way they are able to sustain life in an unusual environment. Moreover, the properties (such as the oxidation state) of their environment depend on the outer, sunlit, world of life. We have much to learn from the Bushmen, and so too can we learn from the hydrothermal communities.

A more likely setting for the first life is in hydrothermal systems that were in shallow water, or exposed on land (sub-aerial) – exactly what the *Challenger* scientists observed. There are several reasons for this. First, sub-aerial hydrothermal systems often have neutral to mildly alkaline chemistry (their pH, which is a measure of acidity, with values below 7 being acidic, ranges typically from about 6 to 8.5), in contrast to the much more acid fluids of the mid-ocean ridge systems (pH about 3.5). Most chemical reactions in living organisms occur at near-neutral pH, and in those bacteria that live in very acid pools, the internal pH within the organism is much less acidic than in the pool. By implication, the first organic synthesis in the first life probably also took place in a near-neutral setting. Furthermore, if the RNA-first hypothesis is correct, one would expect the first life to have existed in a setting where natural fluctuations could promote the detachment and action of ribozymes: experimentally, this needs very mildly alkaline conditions and temperatures around 35–40°C, in the typical range of sub-aerial hydrothermal systems.

Sub-aerial hydrothermal systems may have also provided the necessary mineral catalysts to promote the first reactions, such as clay minerals, zeolites and iron oxide hydroxides, all made in hydrothermal settings. Sub-aerial hydrothermal systems offer three interfaces at which catalysts could have acted to enable the first synthesis to proceed: these are the solid/fluid, solid/fluid/vapour and solid/vapour boundaries. At one part of a hydrothermal system, for instance, reactions could be proceeding in a fluid (e.g. a hot brine). Elsewhere, a cavity produced by a hydrothermal explosion cavity could be filled with vapour dominated by carbon dioxide. Hydrothermal systems change rapidly, and complex syntheses could proceed in systems changing from fluid-dominated to vapour-dominated and back again, in the presence of an assortment of catalysts.

Hydrothermal settings could also have provided the needed containment for the first life. A prospective universal ancestor molecule in the open ocean would soon be isolated from the rare organic chemicals it needed to reproduce itself, and would be broken up before it could replicate. If it made accessory enzymes, they would immediately be washed away and lost. In contrast, a first replicator trapped in some shallow-level hydrothermal cavity would maintain contact with its daughter products, which could be of use to it. In the hydrothermal systems there would be a supply of essential chemicals such as phosphorus, which is mobilized and rapidly redeposited in hydrothermal systems, especially in some types of lava. Phosphorus is crucial to life – the birthplace must have had a supply of it. Magnesium and other metal ions, such as lead, may also have been important in the first putting-together, and are present in hydrothermal systems, as are methane and ammonia. Modern Icelandic geothermal systems can have methane and ammonia in abundance: in contrast, the early atmosphere or sea would have had virtually no methane or ammonia.

Finally, in the hydrothermal setting there is an obvious source of energy to drive the reactions. Nowadays most life depends on photosynthesis. However, in hydrothermal systems there are always marked chemical gradients and fluctuations as cool fluids meet hot fluids, or hot fluids reach the surface. In this fluctuating setting, life could have been first nourished: our ancestors would not have needed to invent photosynthesis before they could replicate.

OTHER IDEAS

A strong case has been made out above in favour of the notion that life began in a hydrothermal system. This is not the only possibility, and we may never know the answer. The other possibilities include the notion of a 'cold little pond', where some of the synthesis reactions would be more easily carried out. Perhaps a tidal mud flat may have been our birthplace – in this setting the flux of fluid, as the sea and rainwater or groundwater from the land came and went, may have provided the energy source for the first living community. Yet another possibility is some extraordinary combination, such as a volcano in a cold or glacial setting, which would provide both hot and cold little ponds. The list is long.

All theories of the origin of life have flaws: the answer eludes us. Currently, the most popular idea is probably the RNA-first hypothesis. But in science popularity does not mean correctness. Protein may have been first, or even some sort of resurrected soup. We do not know.

AN OPINION

To continue with opinion (and it is only opinion, and not even popular opinion at that), it is plausible that life could have been born in a sub-aerial hydrothermal system, in which RNA was synthesized. In this setting, RNA with ribozymes managed to create an RNA-world, a tiny connected community of vesicles and cavities in a hydrothermal system populated by a small number of RNA molecules. Natural selection may then have produced lengths of RNA able to bind proteins and to use other chemicals, such as those which biologists call, in shorthand, ATP, ADP, co-enzyme A and NADPH. Proteins could have been made from amino acids, thermally synthesized from chemicals such as methane and ammonia in the hydrothermal system. Proteins are superb catalysts and would then have taken over the function of enzymes from ribozymes, leaving the modern ribozymes as a vestige of an ancestral RNA world. Eventually, DNA would have evolved accidently from RNA. This would have provided a much more reliable store of information, since it has a complementary double strand. The double-stranded DNA then took over as the basic genetic library, with RNA retaining its intermediate role as the carrier of information to synthesize the proteins. All this would provide a self-replicating organism, but we have left it in a cavity in the rock. How did it get out? Release of the contents of the postulated initial RNA-organism would be a disaster, as the system would drift apart in the open ocean.

Perhaps natural selection provides the answer. In a hydrothermal cavity, lined with minerals, once an RNA-world had been established, and enzymes had been invented, the catalytic properties of clay, zeolites and oxides which may have been so useful in the initial stages would become a capricious nuisance. By this stage the random intervention of the minerals would be a problem. Any organism that discovered how to exclude the effects of minerals would be favoured by natural selection. Thus any strand of RNA that surrounded itself with a bag made of chemicals known as phospholipids (in which, again, is phosphorus) to enclose itself and its organs, or enzymes, would be able to isolate itself from the accidents and errors of mineral catalysis. Natural selection would strongly favour the RNA that could enclose itself with such a protective membrane.

With the evolution of a membrane comes the chance of controlling energy supply. The first replicating molecules must have been assembled by the accidents of a fluctuating chemistry, as temperature, acidity and oxidation state varied. But with the development of a cell membrane, generation of the chemical known as ATP could have begun, an essential step in metabolism.

Thus, by invoking mineral catalysis first, then an RNA-world, we pass to an organized system within a containing membrane. Hydrothermal systems erupt and explode, and earthquakes happen – it would be probable that such an organism, scattered by geological activity, could colonize the rock/water interface at the surface, either in a stream or in tidal regions, or on the ocean floor around a volcanic island. The next miracle would have to be the arrival of photosynthesis. But that is a different problem.

This RNA-first, hydrothermal model has been explored in some detail, to show one example of the various scenarios proposed for life's origin. But it must be stressed that the model is simply opinion. It has no factual, experimental basis, and many other equally good competing models have been proposed. The origin of life remains lost in the abyss of the past. It is time to move on to what we do know, to examine the geological record.

FURTHER READING

Bendall, D.S. (ed) 1983. *Evolution from molecules to men*. Cambridge: Cambridge University Press.

Cairns-Smith, A.G. 1982. *Genetic takeover and the mineral origins of life*. Cambridge: Cambridge University Press.

Crick, F.H.C. 1981. *Life itself: its origin and nature*. New York: Simon & Schuster, Touchstone.

Edmonds, J.M. & K. von Damm 1983. Hot springs on the ocean floor. *Scientific American* **248**(4), 78–93.

Fox, S.W. and K. Dose 1977. *Molecular evolution and the origin of life*. New York: Dekker.

Gilbert, W. 1986. The RNA world. *Nature*, **319**, 618.

3

The Archaean planet

THE BEGINNING OF THE GEOLOGICAL RECORD

St John's gospel starts with the words 'In the beginning'. Transliterated from the Greek this reads as 'En Archi', or 'In the Archaean': this English name is taken from the Greek source to describe the beginning of the geological record and the pervasive influence of life on Earth. As yet we have no agreed definition of when the Hadean (the accretionary period), ended and the Archaean began. We could use the beginning of the geological record, at present set at about 4.3 billion years for the oldest minerals, or around four billion years for the oldest rocks, or we could define the onset of the Archaean as the time when the first replicating organisms occupied the seas. Life – including rabbits – breeds very fast in an open environment until it completely exploits the available food supply in whatever ecological niche it occupies. Once a photosynthetic bacterium had evolved, within a handful of years the seas would be filled with bacteria in the same way that Australia, a handful of years after colonization, is now widely occupied by rabbits. The onset of life is the great divide in the history of the planet. When life began, when our most distant ancestor was born, this planet became different, perhaps unique.

The earliest Western Australian mineral crystals, the oldest tangible things yet discovered on Earth, are up to 4.3 billion years old. These crystals are zircon, which is a characteristic mineral of the continents, although it can occur in small quantities in oceanic rocks: the inference from these Australian crystals is that some continental crust existed more than four billion years ago. Most probably the Earth has been a planet of oceans set in with continents for well over four billion years, although many scientists think that the volume of the continents has grown since the earliest days.

The oldest whole rocks yet discovered, four billion years old, are in northwest Canada. Here, and also in Antarctica and in the Isua region of West Greenland, the rocks are 4 to 3.8 billion years old. The Isua belt is particularly informative. It is a geological raft of exceedingly old rocks, almost miraculously preserved amidst the gneisses and granitic rocks making up the Greenland con-

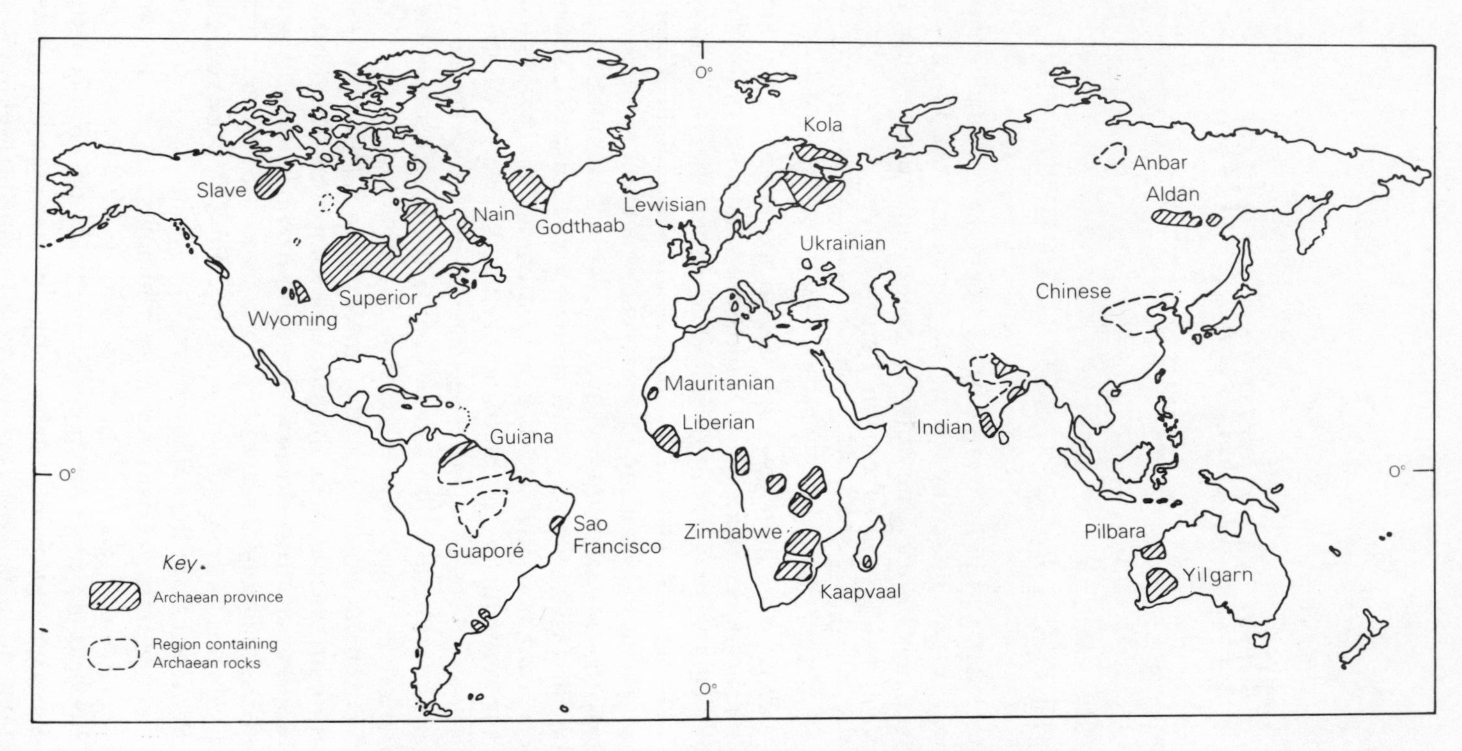

Figure 3.1 Map showing the areas where Archaean rocks outcrop.

Figure 3.2 Ancient gneissic basement, 3.6–2.6 billion years old, south-central Zimbabwe. Several stages in the evolution of the rock can be seen: an old banded rock, folded and deformed like toothpaste; blocks such as that under the hammer; and various stages of cross-veining.

tinental fragment. Rocks within the belt have been dated at about 3.8 billion years old. Over half the belt consists of layered sequences of sedimentary and probably volcanic rock that have been metamorphosed by heating to many hundreds of degrees centigrade and by compression under pressures of several thousand times that of the atmosphere, and have been deformed into complex structures by folding. The sedimentary rocks are varied. Some may have been laid down in shallow water, although there is no necessary implication that a continent, in the modern sense, existed. Nevertheless, these rocks are evidence that land regions stood out of the water, to supply the debris: if the land was not already part of continents as we know them, then the islands that later coalesced to make continents had at least begun to grow.

More detailed evidence comes from rocks about 3.5–3.6 billion years old, in the Barberton Mountain Land, South Africa and the Pilbara block, Western Australia. These rocks are better preserved than the Isua belt, being less deformed and less metamorphosed.

EXAMPLES FROM THE GEOLOGICAL RECORD: THE NATURE OF THE EVIDENCE

The Barberton Mountain Land

In the north-east of South Africa and across the border in Swaziland lies one of the most beautiful and extraordinary parts of Africa, the Barberton Mountain

Land. In summer the green hills of Barberton rise from the granitic plain around: one of the delicate and lovely sights of Africa. For its geological significance this region compares in importance with the celebrated East African localities where fossil ancestors of humanity occur. Botanically, Barberton is also a very special place, filled with rare plants, that deserves to be named as a World Heritage Site. Its geological treasures are now being covered and its botanical rarities are being plundered by collectors or destroyed by pine plantations, yet it remains a stretch of the biosphere that is worth protecting.

Barberton is important geologically for three reasons: the rocks are very old; some are very unusual magnesium-rich lavas, called **komatiites** (after the Komati River which flows around the Barberton Mountains); and, perhaps most important, it contains structures built by bacteria that are some of the oldest convincing evidence for life on Earth. The age of the rocks has been found by a variety of methods, and a figure of roughly 3.5 billion years has been established for some of the volcanic rocks.

Komatiites are lavas, mostly older than 2.5 billion years, which were erupted at very high temperatures, possibly as high as 1600°C (2900°F). Only at such temperatures would these very magnesian lavas be able to flow. Most modern lavas, in contrast, are erupted at rather lower temperatures. The komatiites must have come from a very hot source, deep in the Earth's interior. They were born in an Earth which may have been hotter inside than today. It should be remembered that the Earth's internal temperature and heat loss have virtually

Figure 3.3 Cross-bedded Archaean sand, Barberton belt, South Africa, laid down in shallow water.

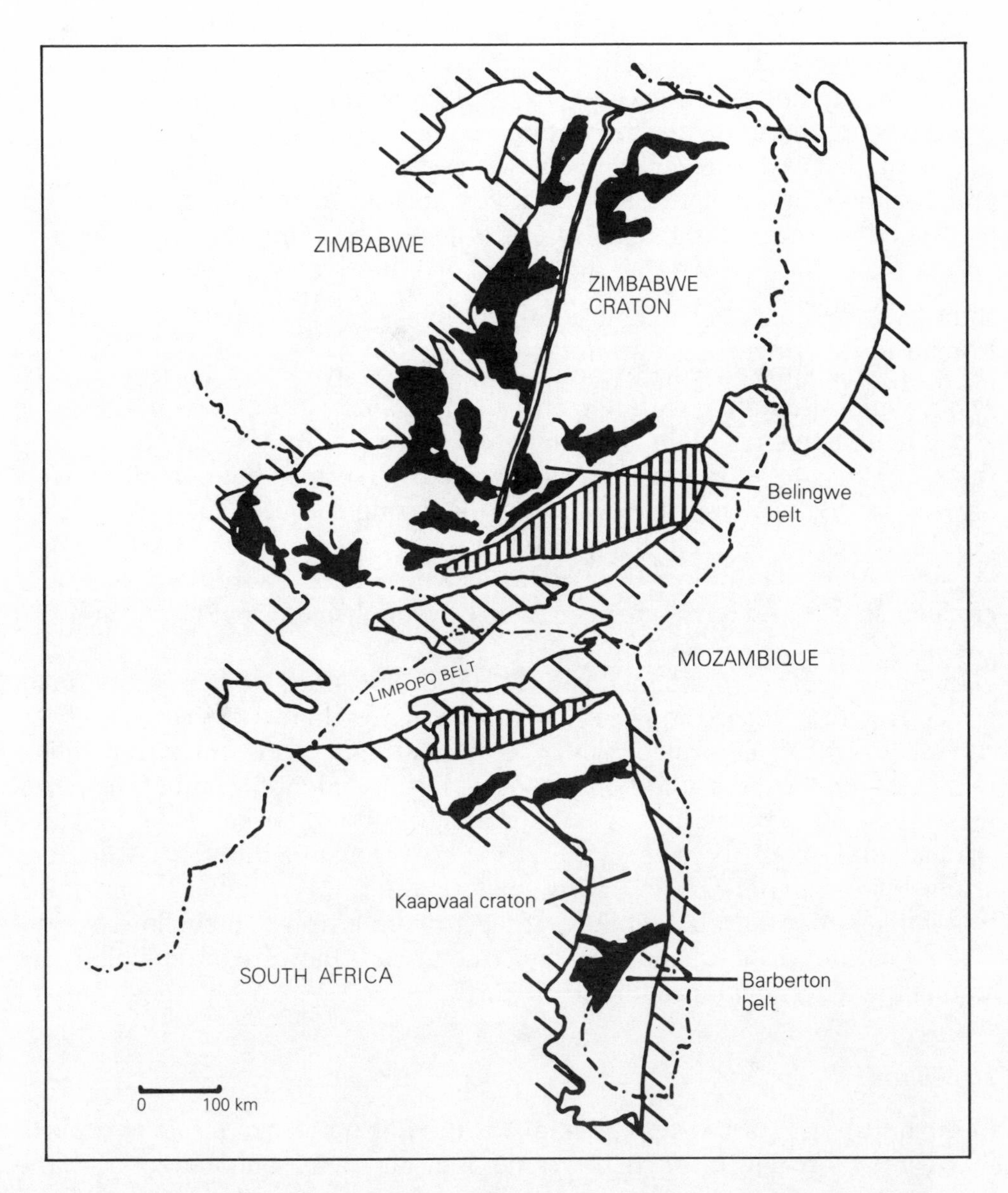

Figure 3.4 Sketch map of some of the important old geological features of southern Africa. Belts of lava and sediment are shown in black. The Barberton Mountain Land contains some of the oldest evidence for life on Earth. It is located on the Kaapvaal craton, which underlies large regions of central South Africa. The Belingwe belt in Zimbabwe contains a wide variety of old rocks, including stromatolite structures built by bacteria, and komatiite lavas erupted at high temperatures. It is part of the Zimbabwe craton, another old continental nucleus. The Limpopo belt, between the Zimbabwe and Kaapvaal cratons, contains some old rocks formed at depths of 30–35 km (roughly 20 miles) at the base of the continental crust. The long thin feature in Zimbabwe is the Great Dyke, an igneous body intruded at the very end of the Archaean. It is a major reserve of chrome and platinum.

Vertical shading shows belts of rock metamorphosed under high temperatures, on the margins of the Limpopo belt. Diagonal shading shows boundaries of regions covered by younger rocks.

no direct influence on surface temperature, which is set by the properties of the atmosphere. The only connection is that internal heat loss produces volcanoes, which vent gas and dust, altering the greenhouse properties of the air. Though komatiites are not common, they are widespread in the Archaean geological record, being found on all the continents. In the Barberton belt they form an important part of the geological sequence, especially in the formations found in and near the valley of the Komati River. In this sequence, komatiites and associated lavas occur together with thin silica-rich sediments called cherts.

Nearby, in the Barberton mountains above the river, in another geological unit, komatiites are associated with quite different sedimentary rocks. In this case, the sediments contain structures that probably originally were **stromatolites**. Stromatolites are rocks formed originally of layer after layer of organic-rich material alternating with calcium carbonate. Individual layers may only be a few millimetres or tenths of millimetres (tenths or hundredths of an inch) thick, but the whole structure may be two metres (six feet) high, or more. They are built by organisms such as cyanobacteria (formerly called blue-green algae). Similar structures are still being formed today, most notably in Shark Bay, Western Australia, and in the Bahamas, where an extraordinary variety of complex shapes and forms are being built by cyanobacteria in tidal and subtidal water.

In the Barberton Mountains a variety of probable stromatolites occurs, small structures preserved in the silica-rich rock known as chert. The rocks have been recrystallized and deformed, with the original minerals of the stromatolites replaced by finely crystalline silica. As a result, the identification of the structures as being stromatolites is not absolutely certain. Some of the probable stromatolites seem to have grown very close to komatiitic lavas. Because stromatolites can only grow in fairly shallow water, where light can penetrate, this implies that these komatiites, at least, were erupted in shallow or tidal water conditions, or subaerially. The structures, if they are stromatolites are among the oldest evidence for life on Earth.

The Pilbara

In the north-west of Western Australia is the Pilbara block, a tract of country that contains rocks of much the same age (about 3.5 billion years) as the Barberton belt. In comparison to the compact Barberton terrain, the Pilbara sprawls over an area comparable to the United Kingdom – one local government unit, charmingly called a shire, is alone the size of a small European country. The landscape evokes images not of Tolkien's shire, but of Mordor: a challenging barrenness, broken by fragments of life. Both Barberton and the Pilbara have long histories of gold mining; both have komatiites and both have probable stromatolites.

The komatiites of the Pilbara are mostly in the Western Pilbara. They are of

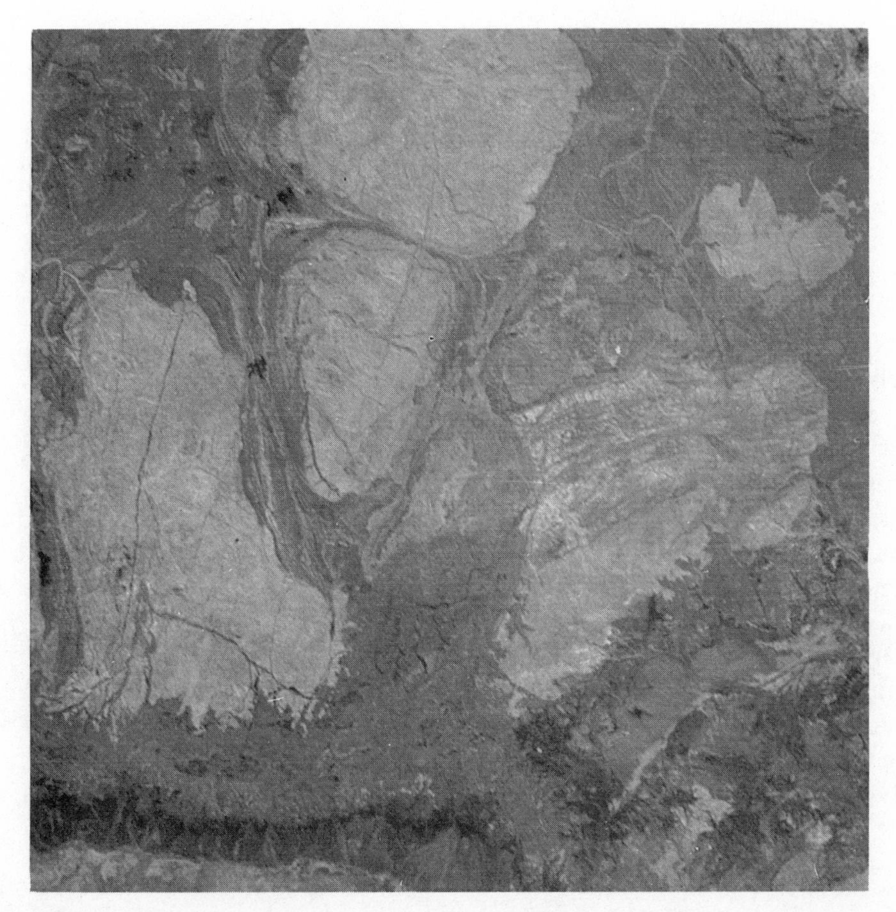

Figure 3.5 Part of the Pilbara region, Western Australia, seen from space. Light-coloured regions are granitic or gneissic; darker coloured regions are belts of metamorphosed lavas and sediments. Much of this terrain is 3–3.6 billion years old. About 160 km, 100 miles across image, North to top (Landsat image courtesy of NASA).

Archaean age, though the exact date is uncertain. The suite of rocks includes komatiite lavas and thin silica-rich rocks that may have been deposited in deep water. Some of the komatiites display very characteristic textures made of plates of the mineral olivine, sometimes up to 60 cm (two feet) long, now completely altered to another mineral, serpentine. This texture, known as **spinifex** texture from the name of the spiky local grass, is typical of komatiites and has also been found in Barberton, Zimbabwe, Canada, and many other places. It is possible that the ocean floor in the early Archaean may have been made from magmas of komatiitic composition, and these rocks in the West Pilbara possibly preserve for us an image of what the Archaean ocean floor was like.

In the Eastern Pilbara, around the old gold mining town of Marble Bar, a very different succession is found. The rocks, which are around 3.5 billion years old, include some lavas but are mainly sedimentary. The geological features of the rocks show that many of the sediments were laid down in very shallow water. Near the locality known as North Pole (ironically one of the hottest places in Australia), some structures occur that may be stromatolites. These rocks, like those in Barberton, have been recrystallized since they were formed: the identification is not absolutely certain. Nevertheless, this is considered to be some of the oldest probable evidence for life on Earth, comparable in age to the Barberton stromatolites.

Another interesting feature of the geological succession in the area is that some nearby rocks seem to have been formed originally as barium sulphate which precipitated when water evaporated. These rocks are known as evaporites. Sulphate evaporites, if they existed, could not have been formed in very reducing (oxygen-poor) environments. This implies that the surface was probably at least neutral to mildly oxidizing by 3.5 billion years ago. A final point about the Eastern Pilbara succession is that it seems to have been laid down on top of a continent, although the contact between the preserved sediments and the original underlying continental surface is not now clear.

These fragments of the geological record, 3.5 billion years old, seem to show that life existed in shallow and tidal waters, that the Earth's interior was hot enough to produce komatiites, and perhaps that some continental areas existed. It should be stressed that much of this inference is based on the interpretation of altered rocks, with complex and difficult textures: our picture of the early Archaean has been laboriously constructed from a small collection of uncertain evidence.

Younger African Archaean successions

THE PONGOLA BELT

In southern Africa, not far from the Barberton Mountain Land, are rocks known as the Pongola Supergroup. This suite, which is nearly three billion years old, consists of some volcanic rocks and a great deal of sedimentary rock: it is rare and special because, despite its great age, it is still little deformed and the rocks are nearly flat-lying. It also contains clear evidence for the existence of exposed land. The rock rests on an older granitic continental surface. Geologists call such a surface a basement, which is the older foundation that underlies the rocks being studied. At the contact, there is a fossil soil horizon made of altered material weathered from granite rocks. In this area, old stable continent has existed with little deformation for nearly 3000 million years. On it, in shallow water, life was present: stromatolites occur.

THE BELINGWE BELT

In south-central Zimbabwe lies the Belingwe Greenstone Belt. The name Belingwe is derived from Mberengwa, which is the local name of the large mountain at the south end of the region. The meaning, roughly, is a place bearing great weight, or carrying great expectation. Belingwe is the transliteration of this name from the original Shona, via the Ndebele language, into English and was the old name for the town of Mberengwa: it is an appropriate name for a stretch of country that contains some of the best preserved of all evidence for Archaean life and the Archaean environment. 'Greenstone' refers to the typical colour of the lavas and sediments.

Figure 3.6 The contact at the base of the 2.7 billion-year-old lavas and sediments of the Belingwe Belt. To the left the rock is weathered, granitic and about 3.5 billion years old. On this basement, the 2.7 billion-year-old sediment to the right – quartz fragments and sands – was laid down. Hammer-head lies on contact which was originally horizontal but has been tilted so that it now dips steeply; (photo courtesy of M.J. Bickle & A. Martin).

As Barberton is special, so is Belingwe. It contains younger rocks than Barberton, mostly around 2.7 to 2.9 billion years old, but they are in a better state of preservation and some are almost unaltered despite their long history. They include some of the freshest komatiites known, as well as extremely well preserved outcrops of stromatolites, all laid down on clearly exposed continental crust. The rocks are set in a country of small sharp hills, leading to the wild mountain clothed by a magnificent forest that is still inhabited by animals such as sable antelope, although the trees are now being cut.

The geological record in the Belingwe Belt stretches over a thousand million years. The older rocks surrounding the belt are around 3.5–3.6 billion years old, themselves incorporating remnants of even older rocks. These old rocks formed an ancient continental basement, which by about 2.9 billion years ago served as the floor onto and into which a series of silica- and magnesium-rich igneous rocks was erupted and intruded. The whole succession was then gently folded and eroded. Around 2.7 billion years ago the process began again, with deposition of shallow-water sediment and then volcanic rocks across the whole terrain of ancient basement and the previously deposited suite of lavas and sediments. The contact, known as an unconformity, between the ancient land surface and the overlying 2.7 billion-year-old material has been preserved for us in several places. It demonstrates clearly that the younger, 2.7 billion-year-old sediments were laid down on ancient continental crust. The continents were thus well established, at least here, by 2.7 billion years ago.

Above the contact are thick shallow-water sediments, including sands and silts showing a wide variety of structures such as ripple marks, which are typical of shallow settings. Associated with these rocks are limestones, many or all of which may have been originally precipitated in mats of algal growth. Some of the limestones contain stromatolites. The geological succession immediately above the sedimentary rocks has been disrupted somewhat, but overlying this are komatiitic and basaltic lavas, many of which have been

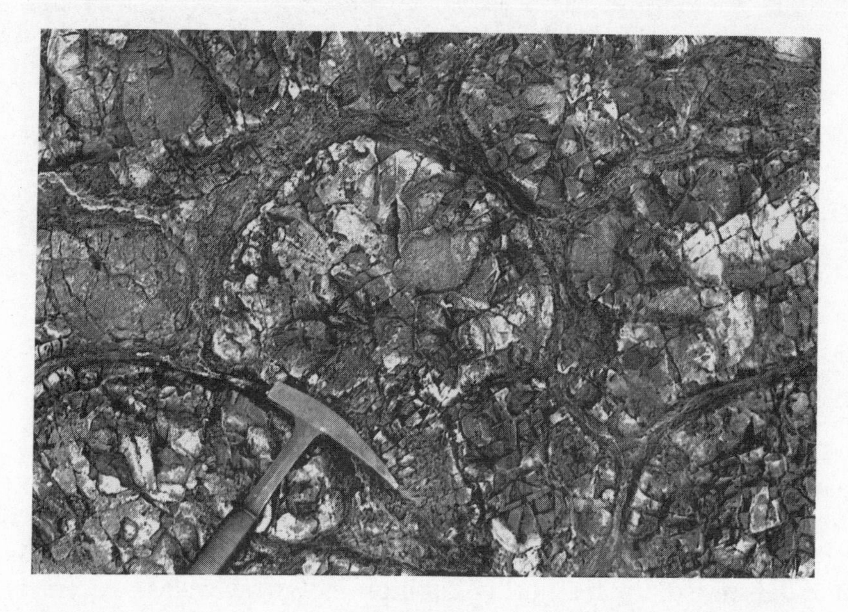

Figure 3.7 2.7 billion-year-old lavas in the Belingwe Belt. The lavas are komatiites and the round structures, known as 'pillows', must have formed underwater (photo courtesy of M.J. Bickle).

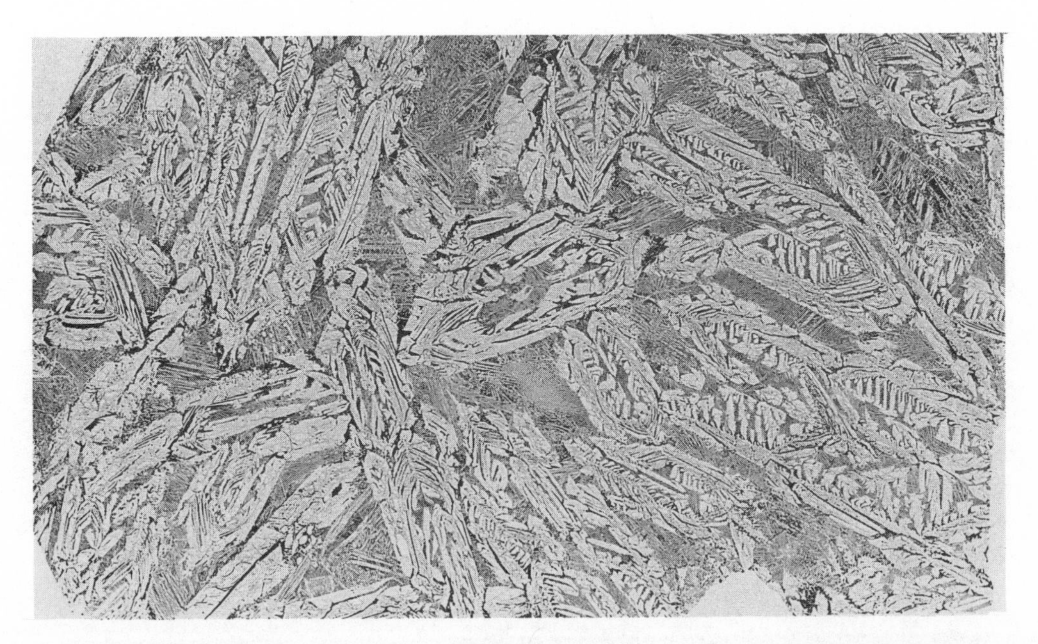

Figure 3.8 Crystals, altered now, but once of the mineral olivine, in komatiite lava from the Belingwe Belt. Roughly 5 cm (2 inches) across photo.

Figure 3.9 Komatiite lava, 2.7 billion-year-old, Belingwe Belt, Zimbabwe. The dark background is altered glass, the crystals are of olivine, which grew rapidly and was then quenched as the lava flow cooled. About 3.25 mm (one eighth of an inch) across the photograph (photo courtesy of W.E. Cameron).

altered, possibly by early hydrothermal systems circulating water around the volcanos. Some of the komatiites, however, possess wholly unaltered minerals and even original volcanic glass (most glass in old volcanic rocks has long since been altered, just as old bottles change colour and crumble). The volcanic rocks are up to 5 km (3–4 miles) thick.

Overlying the volcanic rocks are about 2 km (about a mile) of further shallow-water sedimentary rocks. These too have ripple marks and some show mud-cracks, implying that they were deposited in tidal settings or lagoons which periodically dried up. The most interesting rocks in this succession are spectacularly well developed stromatolites. When examined under the microscope, the stromatolitic limestones are seen to be made of carbonate and wisps of organic material, containing a variety of relict organic chemicals. The rocks seem to have been recrystallized little since deposition, if at all. They present a record of the doings of life, 2.7 billion years ago.

Figure 3.10 Ripple-marked siltstone, laid down 2.7 billion years ago, Belingwe Belt, Zimbabwe (photo courtesy of A. Martin).

THE ARCHAEAN OF ONTARIO, CANADA

In northwestern Ontario, near the little town of Atikokan, is a suite of rocks remarkably similar to the Belingwe succession. The rocks are exposed in the huge open pit of the abandoned Steep Rock iron mine. In the Steep Rock exposures the basement is granitic rock about three billion years old and, as at Belingwe, the ancient land surface is exposed. It has been overlain by shallow-water sediments that include a wide variety of stromatolites, ranging from small structures a few centimetres across to giant examples 5 metres long or more. Above these stromatolites are volcanic rocks, including fragmented rocks produced by explosions.

Much further to the east, near the Ontario/Quebec border, is the Abitibi belt, which contains some superbly preserved volcanic rocks 2.7 billion years old. Among the most interesting of these are the komatiites of Munro Township. Some of the best textures in these rocks have now been destroyed by marauding bands of geology students and collectors, but even so the Munro Township outcrops remain impressive, with enormous crystals of what once was olivine hanging from the roofs of ancient lava flows, now altered and turned from horizontal to vertical dips by the effects of folding.

THE PHYSICAL ENVIRONMENT

Not much remains from so long ago, but there *is* information, and we *can* interpret it. What can these rocks, and the other similarly informative areas tell us about the state of the Earth in the Archaean? The answer is that we can learn much, if we read the rocks carefully and draw comparison with appropriate modern examples. First, there is clear evidence that at least some continents of granitic material already existed in the early Archaean. They floated like rafts on the denser viscous interior of the Earth. We can infer this from the very old zircon crystals in Australia and from the field evidence of sedimentary rocks of Isua, Barberton and the Pilbara, many of which seem to have been derived from a continental source. Stronger field evidence comes from the Pongola, Belingwe and Steep Rock successions which actually preserve old continental basement and the younger material laid down on top of it. More generally, many granitic rocks have been dated at more than three billion years old, and it is these granitic rocks that are the remains of the ancient continents.

We can also guess a little about the thickness of the continental crust. Many of the older Archaean terrains have metamorphic rocks in them, some of which were produced by the effects of heat and pressure on silica-rich continental material. The mineral assemblages in a few of these rocks record pressures that imply a cover as thick as 35 km (or 20 miles) of rock. This cover must once have lain above the rocks now exposed by erosion on the continental surface. Yet today's surface rocks still have thick continent beneath them: in some places,

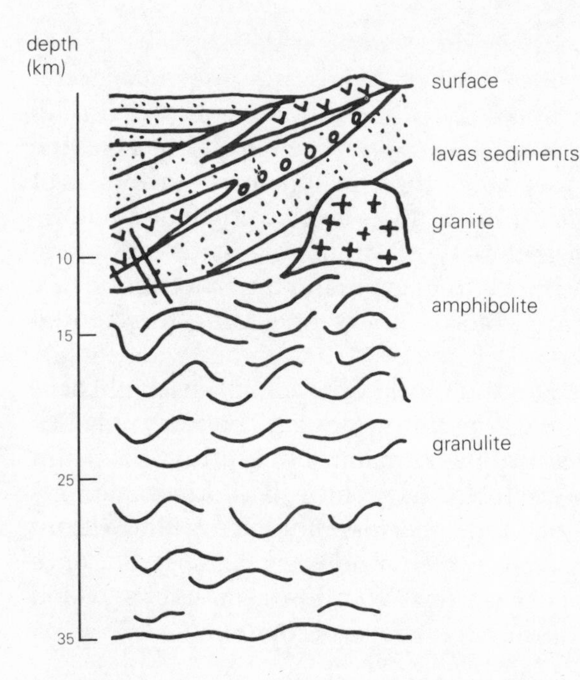

Figure 3.11 A rough cross-section through an Archaean continent. The continents are hugh rafts of relatively light silica-rich rock, resting on the viscous mantle. The upper continental material includes lavas and sediments, often intruded by granitic rocks. At moderate depth, amphibolite grade rocks may occur. These are metamorphic rocks. At greater depth, granulites occur. They are metamorphic rocks of higher density and typically low water content.

such as in West Greenland, the original continent may have been up to 60–70 km (approx. 40 miles) thick, compared with the modern average of 35 km (about 20 miles). This was probably not the general case, because continents then, as now, probably had locally thick regions, but on average the crust of the Archaean continents may have been as thick as in today's continents, or thicker.

The debate about the thickness of Archaean continents is interesting because the continental thickness is determined by the height of sea level and hence by the depth of the oceans. Perhaps the Archaean oceans were on average substantially deeper, say 5–7 km (about 4 miles) deep then, in contrast to 3.5 km (about 2 miles) deep today.

A different problem lies in determining the thickness of the continental **lithosphere**, as opposed to the **crust**. The continental crust is the light granitic raft and all that lies upon it and is mixed in with it, while the lithosphere also includes the cool and mechanically strong material beneath the crust, which forms the bulk of the continental plate. The lithosphere under continents is usually fairly thick, because continents are old and thus have a long time to cool. As they cool they become mechanically strong, just as ice cream is stronger when it has just come out of the freezer than when it has been in the back of a hot car parked outside a supermarket. If we can determine how thick the continental lithosphere is under any particular piece of continent, we can roughly guess the maximum age of that continental fragment. Modern continental lithosphere probably varies greatly in thickness, from 100 km (60 miles) or less in young active areas, to several hundred kilometres in older areas.

Our best evidence for the thickness of old continental lithosphere comes from diamonds. Diamonds can be formed only under high pressures and temperatures, and with the sorts of temperatures one would expect in Archaean continents, diamonds could have formed only at the bottom of the continental lithosphere at pressures exerted by about 150 km (100 miles) of rock, and at around 1000–1150°C (1800–2100°F). The dating of radioactive decay in inclusions in diamonds has shown that many are very old, often over three billion years, although the pipes in which the diamonds were erupted to the surface may be comparatively young. This evidence for the age of the diamonds is corroborated by the existence of diamonds in the Witwatersrand gold-fields, which are probably late Archaean in age (around 2500 million years). These Witwatersrand diamonds now lie in sedimentary rocks and must have been eroded and then redeposited in the late Archaean. Other diamonds – for instance, from Arkansas – are also exceedingly old. The implication of all this is that the Archaean lithosphere was in places at least 100–150 km thick (about 65–100 miles).

For the continental lithosphere to have been so thick more than three billion years ago, it must have been cooling for a long time previously, perhaps 500 million years or more. Therefore, in South Africa we have an example of a continent, part of which may be close to four billion years old. Of course, much continental material is substantially younger – we know from the study of radioactive decay in sediments that the mean age of the continents is around 2000 million years – but there is strong evidence that distinct continents and deep oceans existed in the early Archaean. How extensive the continents were, we do not yet really know.

What of the ocean floor and the interior of the Earth? To study this we must look at the komatiites. Komatiites provide us with a window to the Earth's interior, although a rather dirty window as many komatiites were contaminated as they ascended through the crust and erupted, and most have been altered further since they reached the surface. One thing we can infer from komatiites is that the Earth's interior may have been somewhat hotter than it is today. How much hotter depends on what we can learn about the magnesium content of the lavas: modern oceanic lavas may have 8–10% magnesium oxide (MgO), representing eruption temperatures below about 1280°C (2300°F). In contrast, Archaean komatiites range from 18% to perhaps 28% MgO or more. To be liquid, such lavas must have erupted at high temperatures, up to 1600°C (2900°F). On this rather thin foundation we can build an edifice of speculation.

First, there is the question of whether plate tectonics operated on the Archaean Earth as it does today. Modern plates are driven, in part at least, by the existence of old oceanic lithosphere that is cold and therefore dense, as a result of contraction. Because they are dense, the old oceanic plates eventually fall back into the mantle at subduction zones. Of course, not all subduction zones consume old oceanic plate, but this descent of old, cold oceanic lithosphere is a major driving force of the whole plate system. In this lies a complex paradox: Earth is today living on ancient heat, as it is losing much more heat from the surface than it gains from radioactive decay. By implication, there was more heat in the ancient Earth. If the interior of the Archaean Earth were hotter, it would also probably have lost heat to the surface rather more quickly than today. To lose this heat, the Archaean Earth must have created and destroyed plates more rapidly than the modern Earth. Rapid creation and destruction would have meant that the plates were fast-moving or small, and therefore young when they were consumed. But if they were young and hot they would not be dense enough to fall into the mantle. This is especially the case if the plates were capped by light, altered basalt lava. If so, plate tectonics would have had grave difficulty in operating – to lose heat the plate system would have to work rapidly, but if it worked rapidly there would be little of the old, cold lithosphere that is needed to drive the system by falling back into the interior.

Komatiites offer a way out of this problem (though not the only way). They are denser than basalts. Perhaps the oceanic ridges erupted komatiite, not basalt as they do today. If so, the young oceanic plate would be intrinsically heavier than it is today. More important, just as oil becomes more runny, less viscous, when it is hot, the Earth's interior under a komatiitic ocean ridge would have been less viscous, allowing the plates to move more easily. In consequence, plate tectonics could work even though the average age of the plates was much less. Plates could turn over more frequently, either because spreading rates were high or because there were far more oceanic ridges spreading at today's rates but creating more small plates. In a komatiitic system, therefore, plates could have lost more heat than today; furthermore, more heat would have been lost at komatiitic mid-ocean ridges, as the lavas would have erupted at higher temperatures. The mid-ocean ridge hydrothermal systems would have been more active, cycling the water faster than on the modern Earth. This would have imposed a strong control on the chemistry of the seawater, since the amounts of salt, calcium, magnesium, phosphorus and many other chemicals in the water depend partly or wholly on the rate of hydrothermal cycling. Overall though, komatiitic lavas are not greatly different from modern basaltic lavas, and so the composition of the sea must have been close to that of modern oceans. This is not the only possible solution to the Archaean heat puzzle. It is also possible that the system worked faster, much faster, than today, but produced basaltic, not komatiitic lavas, giving an Earth surface that was more active than today, but of similar composition.

There are other puzzles. For instance, the land area and rainfall have probably changed since the Archaean, and so continental runoff may have been different, influencing sea chemistry. We do not know, but most probably the sea was salt, as sodium is rapidly mobilized in hydrothermal systems. The sea was probably of much the same acidity as today. In some places seawater dried up. When seawater evaporates, it leaves behind its evaporated salts, and there are a few Archaean evaporitic sediments. Some seem to have once been made up of gypsum (a common modern evaporitic mineral), others have more exotic components such as the barium sulphate found in the Pilbara and, possibly, magnesium salts. Overall, though, we would recognize our sea – it was not a bilious yellow mixture of sulphuric acid and assorted horrors thrown in by Macbeth's witches.

The witches might, however, have greatly enjoyed another concoction dreamed up by geologists. This is the idea of a buried magma ocean, which, if it existed, may have been a source of some of the komatiite liquids. At considerable depths in the mantle, around 250 km (150 miles) or more, komatiitic liquid would have been more dense than the mineral olivine that precipitates from it and which makes up the bulk of the upper mantle. In this setting, a komatiitic liquid would not rise: the mineral crystals of olivine would instead float to the top. Any melt in this part of the upper mantle would collect in a magma shell, bounded on top by floating crystals of olivine and below by the denser minerals which exist at great pressure. It is possible that a liquid shell existed around the Earth, with the solid mantle and the lithospheric plates floating on top of it. This shell has been called a LLLAMA, or large laterally linked Archaean magma accumulation. The model is very speculative and has been disputed by Ogden Nash who remarked that

> 'The one-L lama, he's a priest,
> The two-L llama, he's a beast,
> and I will bet a silk pajama
> that there ain't no three-L lllama.'

Perhaps the idea is not as outrageous as it seems – we know that the modern Earth has a liquid shell in its core. The early Archaean planet might have been much more extensively liquid.

Komatiites also provide us with some fascinating evidence which may help us eventually to work out how fast the continents grew. Radioactive isotopes of elements such as potassium, rubidium and samarium decay to argon, strontium and neodymium respectively. When melting occurs in the mantle to produce a magma, the elements are separated. As a result, over billions of years the mantle has become different isotopically from the crust. This difference allows us to estimate the rate of growth of the continents. Komatiites are ideal for this research, because they must have been erupted from the mantle. The evidence is very difficult to interpret, but it does seem that the formation of the continents was well under way by the mid-Archaean. Perhaps all the continental growth took place early in the Earth's history – it is good chemical sense to think that all the continental 'scum' floated to the surface at the very beginning

– or perhaps the continents grew mostly in the Archaean, possibly around 2000–3000 million years ago. As yet there is no scientific agreement on this: the full history of continental growth is still poorly understood. The field evidence, however, shows clearly that by the end of the Archaean much continental crust did exist. The continents have been very extensively reworked ever since by geological processes.

The Earth was ready for habitation, with wide deep oceans and scattered, growing continents. Active volcanoes and hydrothermal circulation around the volcanoes supplied chemical nutrients to the water. Above, a settled Sun provided energy. Life was present. To understand how living organisms developed to occupy every possible habitat, it is necessary first to consider the mechanisms of evolution.

FURTHER READING

Nisbet, E.G. 1987. *The young Earth*. London: Allen & Unwin.
Lewis, R.G. & R. Prinn 1984. *Planets and their atmospheres*. Orlando: Academic Press.

PART 2

The occupation of the planet

4
The variety of life

DISCOVERING LIFE'S HISTORY: STASIS OR CHANGE?

The Earth is filled with life. In the rocks the record shows that life, abundant life, is aeons old. Charles Darwin spent much of his life trying to understand the reasons why life developed as it did. Victorian scientists used drawn out analogies as a way of synthesizing the quantities of data they had hewn out of nature. Darwin compared living creatures to a great tree, on which green and budding twigs represent the existing species. The old wood is the fossil record.

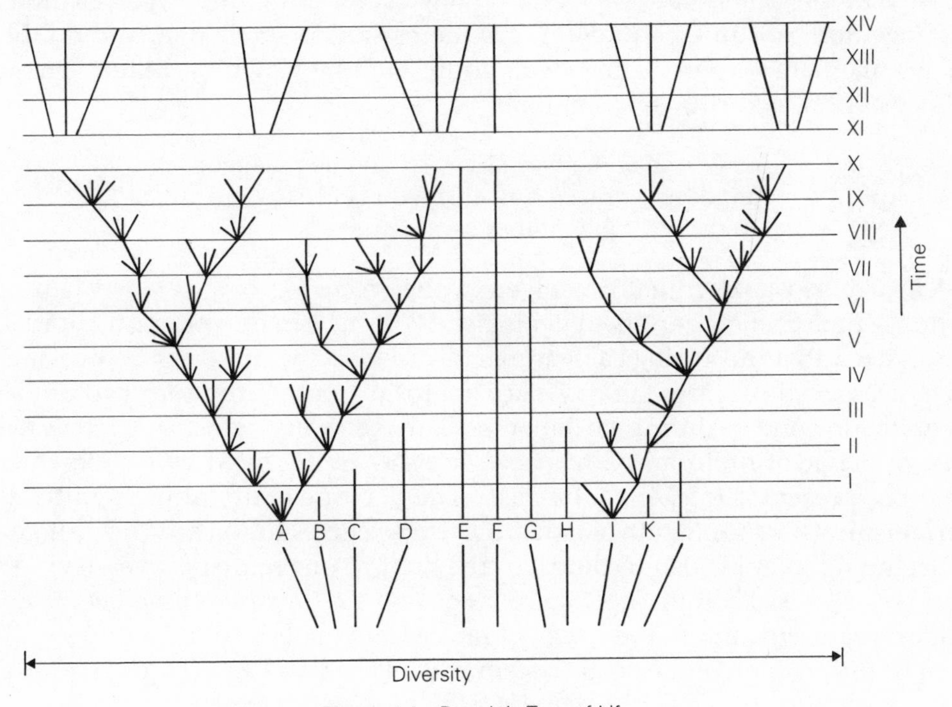

Figure 4.1 Darwin's Tree of Life.

As the tree grows, the living twigs branch out on all sides. Successful limbs shade and kill the surrounding weaker twigs and branches, just as species and groups of species have overcome competing organisms. The great branches of the tree were themselves once budding twigs. Of the many twigs that grew when the tree was a small bush, not many have survived: most have died and dropped off. A few remain, like the platypus, as thin straggly branches. Only a small number have become major limbs of the tree, supporting at the top a dense foliage of new twigs and leaves. Darwin continues:

> 'As buds give rise by growth to fresh buds, and these, if vigorous, branch out and overtop on all sides many a feebler branch, so by generation I believe it has been with the great *Tree of Life*, which fills with its dead and broken branches the crust of the earth, and covers the surface with its ever-branching and beautiful ramifications.' (*The origin of species*, Chapter IV).

The roots of the doctrine of evolution go back to classical times, in a debate between those who saw nature as changeless and those who saw nothing but change and decay. The Roman philosopher, Lucretius, in his work *De Rerum Natura*, wrote 'thus the sum of things is always being renewed' but he also considered that the world is far past its prime: 'the ramparts of this great world will be breached and collapse in crumbling ruin about us'. Shakespeare echoes this sentiment in *The Tempest*. The Elizabethan poet Spenser, on the other hand, in his *Cantos on Mutabilitie*, saw both change and continuity: over all things 'change doth not rule and raigne, but they raigne over change, and do their states maintain'. In his Latin poems (translated by Cowper), Milton agreed. Our parent, the Earth, is not in decay:

> How? – shall the face of Nature then be ploughed
> Into deep wrinkles, and shall years at last
> On the great parent fix a sterile curse?

The debate was eventually resolved by the Scottish geologist James Hutton, who was the first modern field geologist. With his careful empirical approach he showed that the geological features recorded in the rocks can be explained by processes still occurring today, such as erosion, or the eruption of volcanoes, or earth movements during earthquakes. This discovery, which was later called the 'principle of uniformity', has been roughly represented by the statement that 'the present is the key to the past'. The doctrine of uniformity, **uniformitarianism**, is the foundation of modern Earth science. Hutton saw 'no vestige of a beginning', yet he also recognized 'the limit to our retrospective view'. We can now see back, in the geological record on Earth, to the time before four billion years ago; and in the decay of radioactive material in meteorites to the time of the creation of the Solar System, 4.6 billion years ago. Through physics we can see back much further, to the beginning of time itself.

Hutton was followed by Sir Charles Lyell, who had been the student of William Buckland, professor of geology at Oxford. Lyell's book, *Principles of Geology*, published in 1830, had an explosive effect. Across the world it was read by engineers, dilettante gentlemen, university professors, intellectuals of all sorts: it created a discipline of geology. At the high tables of Cambridge, Heidelberg and Harvard, it was the talk of the day. In the dusty hills of Africa it was read by a young road builder, A.G. Bain (of whom more later), who suddenly realized he could read the rocks around him as a book. Most important of all, on the little ship HMS *Beagle*, it went with Charles Darwin on his voyage of exploration through the oceans of time.

The idea of uniformity, as championed by Hutton and Lyell, provided the key to unlock the secrets of time. If uniformity is true, then any feature seen in a rock should have some modern analogue. Once the analogue is identified, the way in which the rock formed can be deduced. On a gross scale, if the rock is a lava, then it must have formed in a volcanic eruption, because that is how lavas form today. On a fine scale, a structure such as a small ripple mark, displayed in a rock, implies that the rock formed from sediment laid down in shallow water, because today most ripples form in those conditions. What is true for the inorganic features of a rock is also true for fossils of organisms. If the shell of a clam is found in a rock, that implies that clams lived at the time the sediment was deposited to form the rock.

More generally, if the pattern of life on the planet today is controlled by certain processes, then those processes must have controlled the past history of life. It was this that Darwin realized. He observed the life of the modern planet, and deduced the general rules by which it changed. Paradoxically, by using the principle of uniformity – what occurs today can be used to explain what took place in the past – Darwin helped to work out the history of life on the planet, and to show how change had taken place. Over time, the Earth has changed, but we can use the present, and the idea of uniformity, to work out how it altered and evolved.

The Royal Navy is chiefly known, despite the lapses of some of its masters, as the instrument that allowed the birth and defence of modern liberal democracy. It also made an enormous scientific contribution to our modern understanding of the world, especially through the voyages of Captain Cook, in the ships HMS *Endeavour*, HMS *Discovery* and HMS *Resolution*. He and his crews travelled as the first scientific expeditions whose aim was to understand our Earth as a whole. They gave us our concept of a planet, alive and interdependent. A century later, the first detailed study of the oceans was carried out by HMS *Challenger*, which sailed around the world mapping the seafloor. Less celebrated voyages such as that of HMS *Enterprise* filled in the details of the map of the planet. These were the space projects of the enlightenment, dedicated to the advancement of humanity. Half a century behind Cook came Charles Darwin and the *Beagle*. Darwin knew Cook's discoveries, and carried with him his geological training. Later, he read an analysis of population growth and competition by the Rev. Thomas Malthus.

Like Cook, Darwin saw the whole Earth, and it was his mind that made the logical steps connecting the diversity of the natural world with the analysis of population. He realized that, when living organisms reproduce naturally, not all the offspring can survive. Rats, for instance, reproduce at an enormous rate: why is the Earth not filled with rats? The reason is that many of the rats die, either from competition between themselves for the available food, or because of competition and predation by other species. The rats that die are the less successful; those that survive pass on their genetic inheritance to the next generation.

Darwin was a cabbage grower. In cabbage he saw enormous natural variation. For instance, he once raised 233 seedling cabbages from some plants of different varieties growing near to each other. Only 78 of the seedlings were true to their kind, and the rest showed variation. All organisms that reproduce show variation; by selecting from this natural variation a breeder can produce change, just as the offspring of the original dogs now range from the chihuahua to the St Bernard.

We have here two separate insights, which together explain the shape of Darwin's 'Tree of Life': there is competition among individuals, and there is variation between individuals. A species is, in Darwin's thinking, made up of a population of individuals, collectively distinct from other organisms, but with each member of the population differing from all others because of its unique genetic heritage. In the competition for survival, only the better-equipped individuals will succeed and reproduce, to pass their genetic heritage down to the next generation. If in a population some individuals have even a slight advantage in reproduction (say 1%), after 100 generations their genetic heritage will dominate the population. This is true however fast the rate of reproduction or however many the offspring of each couple – it applies equally to elephants and to insects. Successful children (in other words, those children that themselves have children) eradicate the genetic heritage of less successful children, successful species eradicate the genetic heritage of less successful species.

It is now known that the genetic heritage of organisms is held within the DNA of each individual. DNA is the information tape of Life: indeed, some radical views on evolution see DNA as the only important factor in evolution, with the organism simply being a throwaway survival capsule for the DNA. When sexual reproduction occurs, the DNA in the nucleus of the first cell of the offspring contains information from both parents. If the offspring is well designed, it will be successful and survive, and its DNA will pass down to another generation; if the offspring is unsuccessful, its DNA will be eliminated from the population. Of course, many other individuals will share common bits and pieces of genetic information, but over the generations the less successful pieces of information will be eliminated. For instance, if among humans a genetic trait such as height over 1.8 metres (or say, six feet), became regarded as so ugly as to be horribly repulsive, tall people would find it difficult to attract mates. They would not pass on their genes. Tall people would still be born, because some tallness genes are probably hidden in most of us, but if sustained over many hundreds of generations the number of tall people would become fewer and fewer. Basketball as we know it would become extinct (or at least the score would be easier to count) unless male and female basketball players decided to intermarry and preserve their race. The population would change: perhaps it would divide into a majority race that was short, and a tall minority race, descended from the intermarrying basketball players.

Selection of this sort explains the details of change, but where is the source of the genetic variation? How did the first organisms produce the incredible variety we see today? Surely all the variation that exists in life today was not present in the first living organism, to be divided up and parcelled out amongst the species?

The answer seems to lie in the accidental changes that can occur in genetic information. The first organisms presumably had rather short information sequences, but genetic doubling can easily occur. Imagine a genetic instruction, meaning 'grow a tail', which goes ABC,DEF. In real life, the genetic code is more complex, but works with similar three-letter words. Now imagine an accident during reproduction that doubles this brief bit of genetic tape: ABC,DEF,ABC,DEF. This accidentally made new instruction will probably be meaningless and the result will be a tail-less individual, which soon dies. Alternatively, the new instruction may produce a monster, such as a fish with two tails. The monster too is likely to die. But there is a slight chance that the new instructions will be useful: perhaps they produce a better tail fin. Furthermore, it might be possible to read the new words as three instructions, . AB,*CDE,FAB,CDE*,F . . , with quite different consequences, especially in the FAB section: perhaps it means grow a second set of fins, or an extra body segment (in reality, things are much more subtle). Next, allow an accident to occur, such as the arrival of a cosmic ray, or simply a mistake in copying during reproduction, to give . AB,*CDE,FFB,CDE*,F . . . Perhaps this change, a mutation, gives yet other new organs. Most mutations will die, but a very few will survive, and occasionally one might be successful.

We know that in many animals and plants (though not in most bacteria), most of the DNA carries information that appears to have nothing to do with the growth or life-processes of the body of the organism. Only a relatively small amount of the genetic information is sense, containing the laws and instructions for building the organism. The rest is nonsense. Human DNA, for instance, contains the information necessary to make a human being, plus much more that appears to be useless. Some animals, such as salamanders, lungfish, frogs and toads carry enormous lengths of un-needed data, so that their DNA contains much more information than ours. Most of these data appear to be what molecular biologists call 'junk' (although some of the junk has a role). DNA can be compared to a cassette tape, and it is as if, when copying the tape to make the next generation, parent organisms do not bother to clean out the noise and rubbish signals, most of which are made during the reproduction process itself. The true signal is interspersed with meaningless passages; later, each time the daughter organism reads the tape, these meaningless passages are simply cut out.

Accidents occur during reproduction: the tape is never copied perfectly. These accidental changes are mutations. Many of the mutations or changes that occur when DNA is reproduced may be 'silent'. In other words, they are not expressed in the body of the organism. Either they are mis-spellings in the junk information that is not used to construct the organism or they are very subtle and make little real impact on the organism. Over millions of years, changes may occur in the DNA with little apparent effect on the body of the living organism, until a final change turns the previous nonsensical and unimportant mis-spellings into sense: a useful mutation occurs. Through mutations come new body modifications and new ways of living. Within the DNA the alterations occur fairly steadily from generation to generation, but they may be expressed only irregularly in the body of the animal or plant. An observer who recorded all the generations would see many frequent small steps in the genetic material, changing the DNA from generation to generation, but only rare and irregularly occurring bodily mutation.

Evolution, then, is a matter of variation and natural selection. The variation occurs by the accidents of genetic transfer, during reproduction in plants and animals. These accidents alter the DNA from generation to generation. Natural selection acts upon the organisms produced by each new DNA tape, allowing advantageous modifications to propagate themselves and deleting the genetic heritage of the less successful organisms.

In his discussion of evolution, Darwin was considering creatures such as animals and plants which reproduce (usually sexually) and die, leaving a new generation to propagate the genetic heritage. In Darwin's thinking, populations could change their genetic heritage only by the success or failure of individuals. The ancestral giraffe did not grow a longer neck because it wished to eat a tree leaf, and it did not pass on that long neck to its children: rather, giraffes with shorter necks starved to death or had fewer offspring and died out, and the longer-necked giraffes came to dominate the population. Furthermore, once a species, or inter-breeding population, is defined, it cannot exchange genetic information with other species. Storks do not produce offspring with a hippopotamus, nor even can a horse mate with a donkey to produce a fertile foal (but see later for the exceptions).

Bacteria, on the other hand, do it differently. They can gain new genetic information from other bacteria during their lifetimes. By exchanging genetic material, bacteria can learn new tricks, such as resistance to antibiotics, without going through the phoenix-like learning processes of higher organisms. A long-necked 'giraffe-bacterium' (if such a thing could exist) could meet another 'giraffe-bacterium' and pass on the ability to grow a long neck: in a sense, all common bacteria can be regarded as a single species, since they can interchange genetic material. This is an ability that the so-called higher organisms – plants and animals – do not usually possess. However, even among the plants and animals there are exceptions, in which genetic material is transferred horizontally across the species barrier. One example of this is the possibility that genetic information in plant tumours, called crown galls, comes from bacteria; another example is a bioluminescent bacterium that has acquired genetic information from a fish. Viruses may be able, on rare occasions, to act as agents that transfer information from one individual to another of the same species. There is also the human immune system, which can learn information throughout the life of the individual. But these are special exceptions: once an animal or plant species is distinct, its evolutionary future depends on what can be done with its own, individual, genetic information. From the interactions between these separate, distinct species has come the modern community of life, collectively constructed by individually competing organisms.

THE INTERDEPENDENCE OF LIFE

Darwin spent much of his time thinking about islands. Imagine a newly erupted, uninhabited volcanic island. A few years after its birth, wind and rain, weathering and dust, produce a few pockets of soil, and windblown or floating seeds colonize them. Birds arrive, bearing seeds on their feet and in their droppings, and a simple community of life – an ecosystem – is established. If the island is very isolated, so that only windblown seeds and birds can immigrate, there will be a limit to, or at least a slowing-down of, the immigration, when all those species that can arrive have arrived. That is not the end of the story, though. The birds will adapt to the trees, and the trees to the birds, by natural selection. For instance, a bird species may slowly lose its wings if there is no need to escape predators, and part of the population may adapt to produce a distinct ground-living species with a seed-eating life. Simultaneously, the plant whose seeds are eaten may gradually evolve a fruit attractive to the birds, so that the birds eat the fruit and then excrete the seed, moist and surrounded by good manure, and far enough away from the parent tree not to compete with it. Eventually a complex ecosystem evolves with many new species, each link interdependent on the others.

This is true not only of islands but of all life. The branches of the tree depend on each other, even though they compete with each other. Predators depend on prey, prey depends on predators. The whole system is in a spiral of dynamic evolution, forever evolving with small changes in the society of life. Each 'improvement' in a lion brings a matching change in zebras and impalas for, if the prey does not adapt, it is extinguished and another prey species takes its place. Overall, there is no equilibrium, but rather a fluctuation about an ever-altering average state. There is constant competition, but there is also total interdependence. This is co-evolution. Occasionally, the system crashes.

THE GAIA HYPOTHESIS

The interdependence of life may, however, go deeper. J.E. Lovelock and Lynn Margulis have suggested that the interdependence of living organisms is so total that Darwin's 'Tree of Life' approximates to a living organism. Their idea, called the Gaia hypothesis, is that it is possible to regard the organic world as a system, in the engineering sense: a complex network of feedbacks and controls which collectively regulates the surface of the planet. A simple household analogue is the control of a domestic heating system. When the temperature of a thermostat falls slightly below the setting, the heat is switched on, until the temperature rises slightly above the control level, when it is switched off again. This is called a cybernetic feedback. The Gaia hypothesis is that it is valid to treat the biosphere as if it were a cybernetic self-regulating system or, in biological English, a physiology, for this is how the physiology of an animal works.

There is no sense of ultimate purpose in this hypothesis, although purpose has been read into it by some non-scientists. The Earth is *not* a sentient being, nor is the biosphere. The hypothesis simply implies that controls have evolved which sustain and regulate the biosphere. For instance, why does the Earth have a surface temperature that allows liquid water to exist? One answer is that it is an accidental consequence of the Earth's position in the Solar System and the composition of the atmosphere. The Gaian answer is, in contrast, that controls have evolved to regulate the Earth's temperature, as the temperature of the human body is regulated, by the net effect of the actions of innumerable plants and animals which collectively control the composition and greenhouse effect of the atmosphere and the colour of the surface. This answer is not anti-Darwinian: it simply demands that selective pressures must exist to favour the survival of organisms of the right colour, or which produce or consume the right greenhouse gases, where 'right' means of the sort which promotes the collective well-being of life. The Gaia hypothesis is therefore almost a 'shorthand' description (like all good physical hypotheses), making generality out of the rules that control the interaction of species. It tells us nothing directly about metaphysics: those rules are the rules of the standard physical universe. They may indeed have been set by a transcendent deity, but Nature is under them, not creating them.

It can be argued that the Gaia hypothesis comes quite naturally out of

Darwinian selection. Imagine a small plant which, by accident, makes some by-product that 'improves' the local environment as far as the plant is concerned, say by changing the colour of the surface, or by emitting chemicals that promote rainfall. This by-product of the plant's metabolism improves the setting around it and the plant is favoured: more plants grow. Now, imagine a community, such as a forest, that includes the plant which accidentally 'improves' its environment. The forest is interdependent: the plant can survive only when surrounded by the by-products, such as shade or dead leaves, from the other trees. If it so happens that the forest improves its regional (as opposed to local) environment, then the forest will be favoured. Organisms do not exist independently in a void; they are interdependent, and something which favours one member of a system may well favour the survival of the whole community; in reverse, removing one part of an ecosystem can damage the whole. So if one part of a community improves the regional environment, the whole will have a better chance of survival.

Now imagine a planet of successful communities. Each local system – such as a rainforest, or a community of marine algae – contributes to the character of its region. Those communities which can work with, rather than against, others in the general region will be favoured and will spread. In contrast, communities that modify the environment against the workings of others will be less favoured. Eventually, the result is an interdependent set of communities, created by natural selection, acting collectively as a cybernetic system.

The Gaia hypothesis is *not* generally accepted by biologists, although it commands much interest and respect from geophysicists and atmospheric chemists. It has had successes, for instance, in predicting sulphur transport by a gas known as DMS. However, it must be stressed, the hypothesis remains a matter of vigorous debate. If it is accepted, it will complete the Darwinian synthesis. Gaia, if true, is Darwin's other face: interdependence coming from competition.

FURTHER READING

Darwin, Charles 1859. *On the origin of species*. Many inexpensive paperback editions.

Loomis, W.F. 1988. *Four billion years: an essay on the evolution of genes and organisms*. Sunderland, MA: Sinauer.

Lovelock, J.E. 1979, reprinted 1987. *Gaia: a new look at life on Earth*. Oxford: Oxford University Press.

5
Life in the Archaean

THE EARLY RECORD OF LIFE

The Earth in the Archaean was a planet with deep oceans, distinct continents, and perhaps some limited continental-shelf sea. What can we discover of the history of life?

The evidence is twofold. First, there is the evidence from the stromatolites, and other fossils or quasi-fossils left in the rocks, telling us something about the environments occupied by life. More subtly, the geological record also contains evidence of the actions of life. Like hydrogen, which comes in three forms, or isotopes, as ordinary hydrogen (mass 1), heavy hydrogen or deuterium (mass 2) and tritium (mass 3), the elements carbon, sulphur and nitrogen come in several isotopes. The isotopes of carbon, sulphur and nitrogen are selectively used in biological reactions. This selectivity leaves a fingerprint on the material processed, which has characteristic ratios between various isotopes, depending on the biological reactions that have taken place. If the material is preserved geologically, the characteristic isotopic ratios can also be preserved; moreover, the entire biosphere is affected by the isotopic consequences of life's activities, and so the record is everywhere. From this fingerprint, left in the geological record, we can estimate when the major biochemical reactions began. In a few places, complex organic chemicals are also preserved which may tell us something about the organisms that produced them.

The second line of evidence about the early history of life lies within ourselves and in every living organism on Earth today. The early Archaean life forms were our ancestors, and there are traces of this ancestry in the genetic make-up of all modern life, just as our surnames may reveal the origins of our forebears. The ancestors of Kennedy and Reagan were Irish, while the ancestors of Carter, Ford and Bush were probably English peasants. In the same way, the genetic makeup of our cells tells us about our origins. Of course, there is complexity too, just as Afro-American slaves and, later, immigrants from Europe entering the US at Ellis Island often acquired English surnames. From the DNA in living organisms we can deduce relationships between species, using this type of information.

To use another analogy, imagine a visitor to Earth from outer space using our languages to investigate our history. The extraterrestrial spy would probably deduce that the languages spoken in the USA, France and Germany are in some way related to each other, but quite different from Chinese and Japanese. With a bit more study our visitor might even deduce that English is an extraordinary chimera, with German roots into which have been incorporated vast amounts of French, followed by the addition, almost *en bloc*, of most of the vocabulary of classical Latin and Greek. Some words have been conserved for an extraordinarily long time. The word 'succession' written in a good italic hand on a piece of papyrus by a Roman soldier in Egypt two thousand years ago is still perfectly readable by any English speaker today. Other words undergo odd changes: for instance, 'school' is derived from a Greek root meaning 'leisure', or 'stopping work', which is when education took place. The Victorian sanitary engineer, Mr Crapper, who built fine water closets labelled with his name has suffered an even worse fate.

The DNA, RNA and proteins of living organisms have complex histories. Some of the genetic information has long been conserved. Other information has changed its function greatly, or has been incorporated from one organism into another. We can use this information to obtain some idea of the succession of life. One of the central discoveries of molecular biology is that all organisms are related. Almost all organisms build their proteins out of the same 20 amino acids, and all organisms use the same code to translate the information stored in nucleic acid into proteins. This code is the **universal genetic code** shared by all life, with only the most minor of exceptions. The implication of this is that all life is descended from a single organism or a very small interbreeding group.

This molecular record gives us some idea of life's evolution; the geological record helps us to describe its habitat.

The geological record – life in the rocks

Modern stromatolites are made by diverse colonies of algae and bacteria, but the closest analogues to the Archaean examples are probably stromatolites made by cyanobacteria, formerly called blue-green algae, such as those in Shark Bay, Western Australia. Most probably the Archaean examples, such as the Belingwe rocks, grew in similar settings, in tidal lagoons where the growing stromatolites were protected from being buried in sediment carried down by streams or by tidal currents. There was probably a diverse community of bacteria at work on the floor of the lagoons and in shallow subtidal waters. Evidence for this diversity comes from the wide variation in stromatolite shapes and forms: some of the variation may be simply environmental, some is probably the result of evolution in the biological make-up of the living community. These changes can be seen in the mineralogy and texture of the rocks, and also in their isotopic record. For instance, samples may show marked differences in their content of the isotopes

Figure 5.1 Stromatolite domes, about three billion years old, Steep Rock belt, North-west Ontario, Canada. The rocks have been tilted, so that the original horizontal now dips about 70° to the left. The stromatolite domes are up to 4–5 m (12–16 ft) long and up to 2 m (6 ft) high. They were built by cyanobacteria.

carbon-13 and carbon-12, or in oxygen-18 and oxygen-16. There are considerable variations in the carbon, oxygen and nitrogen isotope contents of Archaean stromatolitic rocks, and the reasons behind this variation may in part be biological. Isotopes can, in principle, also tell us the temperature of the water in which organisms grew, because the isotopic make-up of organisms and chemically laid-down sediments varies with temperature.

When a bacterium incorporates carbon to its body, the biochemical reactions, such as those in photosynthesis, separate the carbon isotopes slightly, so that the organism has an isotopic ratio that is slightly different from the surroundings. The isotopic composition of the carbon-rich deposit that is eventually produced when the organism dies depends on the nature of the biochemical reactions, on the temperature at which they took place and on the way in which the deposit was preserved. Unfortunately, there are too many variables in the system for us to be able to fix Archaean temperatures with confidence, but one preferred model implies that the temperatures in, for instance, the Archaean Belingwe lagoons were not greatly different from those in modern tropical to

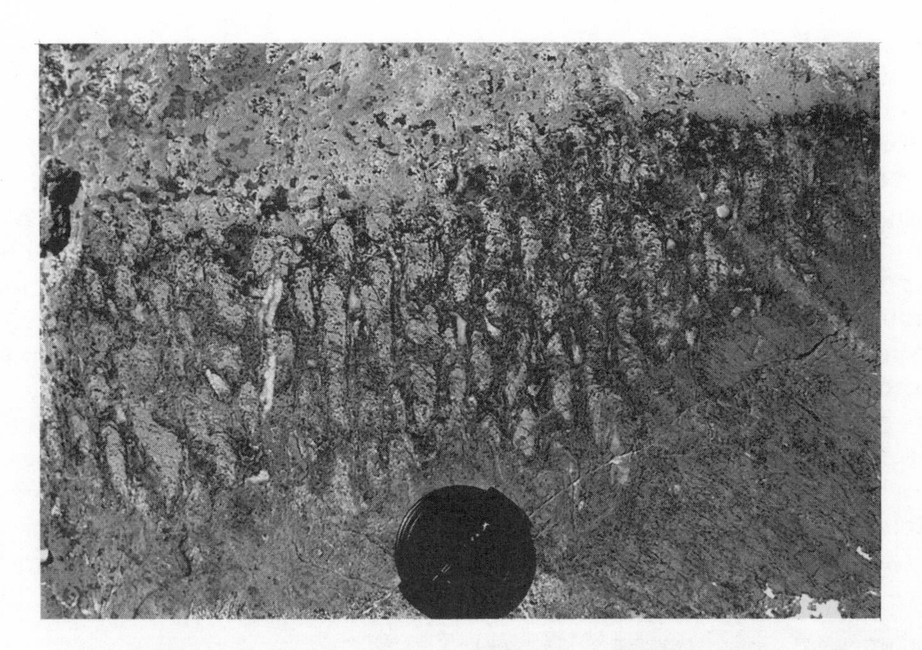

Figure 5.2 Detail of stromatolite columns, also from Steep Rock, Ontario.

Figure 5.3 A well preserved stromatolite horizon in 2.7 billion-year-old rocks, Belingwe, Zimbabwe (photo courtesy of M.J. Bickle).

subtropical settings. Similarly, in some stromatolites the nitrogen isotopes vary greatly, from one lamina in the rock to the next, in a way that seems to be recording changes in the intensity of biological reactions taking place on or just below the water/sediment interface.

We do have another guide to the temperature of the Archaean Earth. In various places, especially in the Witwatersrand sediments containing the South African gold-fields, geologists have described strata that appear to have been deposited in glacial events. If so, temperatures may have occasionally dipped enough to cause ice ages. Unfortunately, we cannot be certain as it is difficult to distinguish between many glacial deposits and the products of sudden sedimentation processes – landslides, catastrophic floods, submarine avalanches and the like. However, one or two ice-scratched boulders may be the clue to the matter: it is likely that Archaean ice ages did occur occasionally. If so, the inference is that the planet's temperature fluctuation was, over the Archaean, much the same as in the last 500 million years.

The geological record tells us that in the Archaean a diverse and important

Figure 5.4 Detail of stromatolite growth, Belingwe belt, Zimbabwe.

living community existed in shallow water, probably at temperatures similar to those on the Earth today. This is exactly what one would expect: life breeds like rabbits, and it would fill the Earth to the limits of its ability, and then compete with itself to diversify that ability. The organisms were probably all bacterial, but they may have been very diverse and extremely abundant. The oldest isotopic record for life seems to come from the 3.8 billion-year-old Isua rocks in Greenland. When one makes allowance for the effects of metamorphism undergone by these rocks, the stable carbon isotopes preserved in the rocks seem to be similar in ratio to modern carbon in comparable settings. Because in modern sedimentary rocks the stable carbon isotope ratio is set by biological processes, the implication is that, even as long ago as 3.8 billion years, life controlled the cycling of carbon through the atmosphere and oceans. This evidence from Isua is tantalizing but is not regarded as conclusive. However, by the time the Belingwe stromatolites were formed, the carbon cycle in the air and ocean was certainly heavily managed by life.

What was the extent of the habitat of this abundant life? There is only a fragmentary record of Archaean shallow-water sediments, especially limestones, which are very rare. We do not know why this is. Possibly the continental area was small, so shelf seas were limited; alternatively, the Archaean Earth was so active that the chance of preservation of a shallow-water platform was low (the Pongola belt, though, is such an exception). California today has very few limestones, but there is nothing very unusual about its biology or geology. It is simply a rather active continental edge: a geological setting not conducive to the deposition of thick limestones. Life occupies every possible niche, even in California. In the Archaean, life must have been widespread, occupying the seas, and probably also in the rivers, lakes and streams. There may even have been bacteria on dry land.

The top 30 m (or 100 feet) or so of the water, where light is abundant, is known as the photic zone. Bacteria can float, and some bacteria are capable of surviving much ultraviolet light. Perhaps they thoroughly populated the oceanic photic zone soon after they first evolved. There is another setting in which early bacterial communities established themselves, as some of the oldest bacterial lineages seem to need such conditions: the hydrothermal systems around mid-ocean ridge volcanoes (and also around volcanoes in island arcs). Modern deep-ocean hydrothermal systems have a complex biological community existing around them, ultimately depending on bacterial processes. Some deposits, which occur in Archaean mineral deposits that are believed to have formed in hydrothermal systems, contain organic chemicals. Perhaps these too are evidence for a hydrothermal community in the Archaean.

We can imagine the Archaean living community as a widely populated bacterial world, say around 3.5 to 2.5 billion years ago. This community was a complex society, each group of bacteria having its own role, and being dependent on the activities of other groups. The bacteria would have included

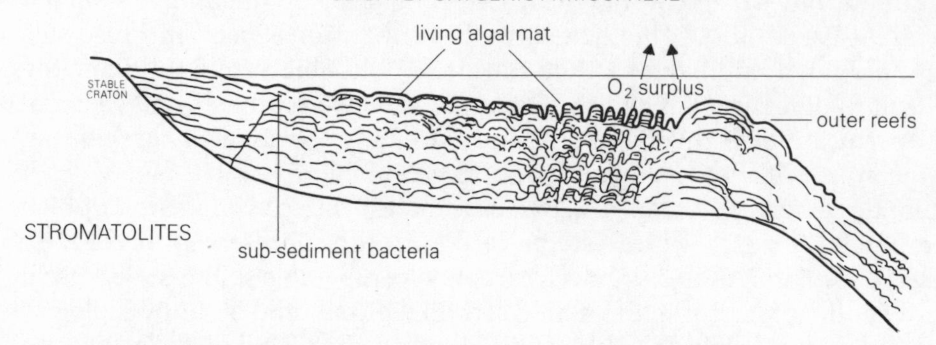

Figure 5.5 Reconstruction of a stromatolite ecosystem. (courtesy of Paul Copper).

cyanobacteria in shallow water, probably photosynthetic bacteria living in the photic zones of the oceans, together with a myriad other organisms living off the byproducts and the debris of the photosynthetic organisms, and a community of bacteria living around the hydrothermal systems of active mid-ocean ridges and volcanoes. There might also have been life on land.

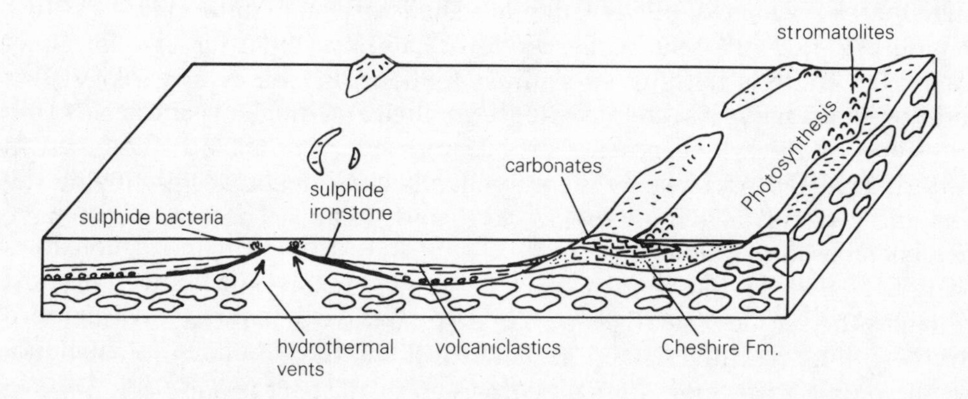

Figure 5.6 A reconstructed stromatolite ecosystem in the Belingwe belt. The underlying rocks are a set of lava flows; the Cheshire Formation is mostly sandstone, siltstone and limestone. Volcaniclastics include volcanic ash and debris, sulphide ironstone is a sediment rich in quartz, iron and sulphide minerals.

The molecular record

The descent of life in the Archaean is best understood by looking at the genetic make-up of living organisms, supplemented by isotopic evidence that may help to date some of the critical steps in evolution. The science of molecular palaeontology dates back to the early days of molecular biology, when it was realized that nucleic acid and protein sequences carry a historical record. The basic starting premise is worth reiterating: all living organisms are related, since all share the same genetic code. All life shares a universal ancestor. Descended from that ancestor, there are distinct groups of living organisms.

How do we divide up life, especially bacterial life? How are the distinct groupings identified? To go back to the example of the spy from outer space, perhaps she would discover that English, German, French, Zulu, Xhosa and Vietnamese all use the same alphabet or code of writing. Even if the visitor did not understand these languages, she might deduce from looking at books in each language that the languages could be grouped into three classes: the European family, the African family and the single Asian language. With more work, but still without understanding the meaning of the words, it would be possible to deduce the transfer of information from one language to another. For instance, technical terms are widely used, in many languages. These terms are usually words that have relatives in the European languages and are obviously derived from them or their classical roots. In contrast, 'impala' is clearly an African word that has been absorbed by English.

Genetic information in organisms is usually analysed either by looking at the sequences of the nucleic acids, or at the sequences in proteins made from the DNA code. Initially, most molecular palaeontology was done by laborious sequencing of amino acids in proteins, but nowadays it is much easier to work directly with the DNA or, usually, with the RNA that is involved in the synthesis of proteins. This work has shown that cellular life can be divided into three groups. There are two groups of bacteria, the **eubacteria** and the **archaebacteria**, all **prokaryotes**, or organisms that do not possess a nucleus to the cell. The third group is the **eukaryotes**, which do possess a nucleus. Some eukaryotes are single cells, others are multi-celled. All plants and animals are eukaryotes, as are fungi and simple organisms called protists.

The two bacterial groupings probably came first. Perhaps there are more than two groups of bacteria – there is some controversy about the point – but most available evidence suggests only two. The eubacteria include most modern bacteria, while the archaebacteria include certain bacteria such as methanogens, which can live in places such as sewage farms and pond bottoms (generating gas), many bacteria living in hot, acid or salty environments, and also many of the bacteria living in hydrothermal systems at mid-ocean ridges, beneath kilometres of water. The two groups of bacteria appear to have diverged separately from a common ancestor, very early on in the Earth's history. Within each main group we can identify distinct lines of descent to subgroupings. The distinction between species of eubacteria is not as clear as in plants and animals, because some widely differing bacteria can occasionally exchange genetic information with each other. However, identifiable groups of organisms do exist, defined by their body-plan and by their ways of carrying out chemical reactions. For instance, the cyanobacteria are an identifiable group of eubacterial organisms that carry out photosynthesis. Although there has probably been a great deal of exchange of genetic information between the

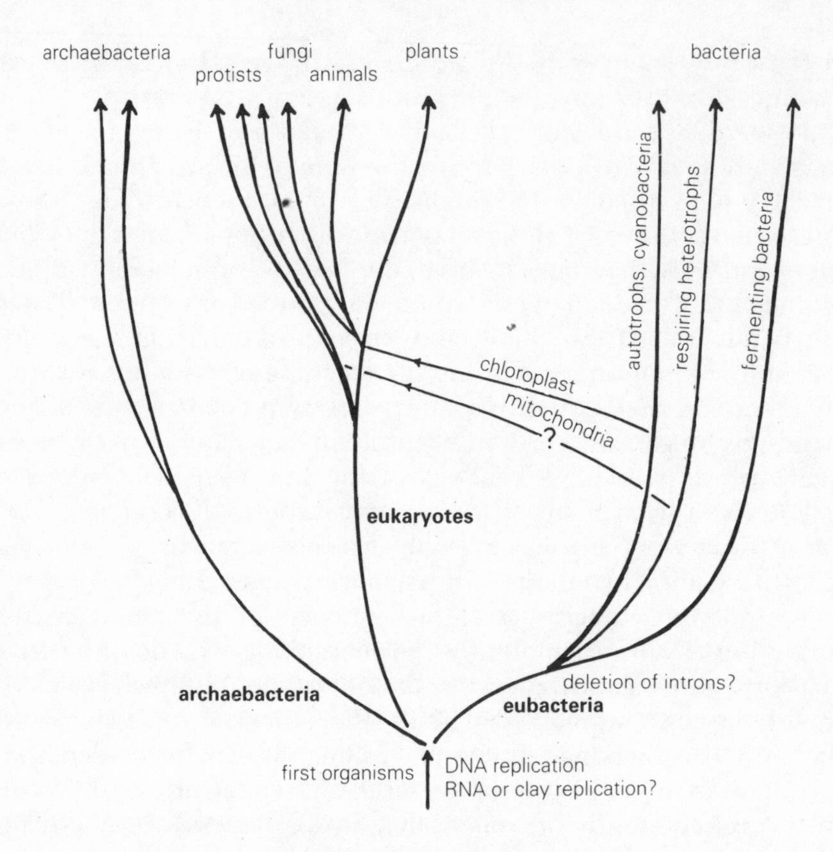

Figure 5.7 The descent of life.

distinct bacterial lines of descent, just as most modern languages share the word 'telephone', the major subgroupings of bacteria are probably very old indeed, and the two main groups, archaebacteria and eubacteria, may have been distinct almost from the beginning of life.

The third division of life, the eukaryotes (which include ourselves) is quite different. Like the English language, the eukaryote cell is a chimera, a Frankenstein's monster, assembled from bits and pieces of genetic information that appear to have been acquired from a variety of sources. Possibly there was an early bacterium-like ancestor, which lived symbiotically with another bacterium or was incompletely eaten by it. At some moment, the partnership became marriage, and one organism resulted. For instance, the **chloroplasts**, which are the parts that carry out photosynthesis in modern plant cells, contain DNA and genetic information that is quite separate from the DNA of the cell nucleus. The chloroplast DNA from modern plants appears to be closely related to the DNA of cyanobacteria (the stromatolite builders). Presumably a single-celled ancestor of modern plants lived closely in association with a cyanobacterium, and eventually the cyanobacterium became physically incorporated into the plant cell, though maintaining some genetic and reproductive identity. Similarly, components known as **mitochondria**, which are possessed by all plant and animal cells, were derived from bacterial symbiotes and carry separate genetic information. In each of our cells we humans depend on mitochondria: we are ourselves symbiotic communities. Some of the eukaryote heritage, therefore, comes from eubacteria. Today some of the symbiotic eubacterial genetic information has actually been incorporated in the nucleus. Other information remains distinct, in the DNA of mitochondria in both plants and animals and of chloroplasts in plants.

Eukaryotes also share important characteristics with the archaebacteria. Earlier it was stated that in many organisms, the DNA tape contains an enormous amount of extraneous genetic information that is apparently useless in the function of the cell, though it may have other roles. Eubacteria do not have these nonsense sequences in their genes, but archaebacteria and eukaryotes do. In eukaryotes, when the tape is processed to make a protein, a segment is copied onto RNA and then the unwanted data, or **introns**, are cut out and the lengths of wanted information are spliced together, rather in the way that a recording engineer will cut out the various discussions between the conductor and the orchestra when splicing together a tape of a symphony after a session in the studio. Both eukaryotes and archaebacteria have introns in their genetic sequence. In contrast, eubacteria are genetically more streamlined and do not generally have introns in their DNA genes that code for proteins, so in eubacteria there is no necessity to splice together the useful lengths of the tape before it can be used.

It is possible that eubacteria never had introns, but it is also possible that introns may be a very ancient feature of life. If so, the eukaryotes and archaebacteria may be preserving something very old indeed, which the eubacteria have lost. Possibly (a pure speculation) this loss happened during early adaptation of life to shallow water environments under bombardment by energetic ultraviolet light from the Sun. In such a dangerous setting, any extra genetic manipulation, such as that necessary to splice together the useful lengths of tape, would incur the risk of random error. Possibly the eubacteria lost the splicing process when their genetically more streamlined ancestors survived and reproduced in shallow water with fewer errors than sister organisms that bore more introns. On the other hand, introns can also be useful in evolution. They may contain junk information, but on occasion it is useful to have a bit of junk lying around. Much of the evolution of the eukaryotes may have been aided by the accidental incorporation of a useful bit of junk into the genetic message.

The driving force of evolution is natural selection – in a population of organisms the fitter will be more likely to survive and pass their genes down to the next generation, but the changes occur on a molecular level, as discussed in the last chapter. The hidden steps in the genetic message from one generation to the next are a sort of molecular clock. We can estimate roughly how long it was since two species diverged by counting the differences in their genetic information. Humans and chimpanzees have almost identical DNA, but are together widely different from cats, and progressively more different from snakes, and fish, and snails, and plants and bacteria. The whole evolutionary tree is written in the DNA of living organisms, and the extent of the differences is a clock that records the time since the ancestors of each have diverged. Of course, analysing the clock is difficult: change may not be a steady process but a sequence of accidents, and clocks based on different parts of the DNA sequence seem to tick at different speeds. Nevertheless, this is one of the most powerful of all palaeontological tools, and when properly calibrated against the fossil record it is the best way we have of discovering the descent of life.

In studying the descent of life in the Archaean, we attempt to calibrate the clocks by identifying the isotopic products of certain bacterial reactions. We suspect that certain sulphur bacteria appeared prior to three billion years ago, and there is evidence that nitrogen was organically controlled 2.7 billion years ago (and probably long before). The evidence as yet is very fragmentary, but it is likely that the major bacterial reactions evolved some time in the earliest Archaean. A complex web of interactions between organisms and the environment seems to have been set up in the Archaean, each main group of bacteria depending on the other groups for its existence.

Perhaps the roots of this ecological web go back almost to the beginning of life. The deepest split we know in the tree of life is that between the eubacteria and the archaebacteria (leaving out the eukaryotes for the moment). If life began in a hydrothermal system, perhaps the archaebacteria, many of which still live in such extreme settings, represent a community that remained mainly in the hydrothermal environment while other bacteria managed to exploit the shallow water environments by means of a variety of new chemical reactions. Photosynthesis, using sunlight for energy (a process of which more is said later), must have come very early in the history of life, especially as the stromatolite record goes back so far. Furthermore, the Archaean hydrothermal community of life would have needed a supply of oxidized chemicals in seawater. In the earliest part of the Earth's history, the oxygen may have been supplied by ultraviolet dissociation of water at the top of the atmosphere and subsequent loss of hydrogen. This source would have declined by the early Archaean, and must have been replaced by the oxidizing effects of photosynthesis. Hydrothermal life is not an isolated system independent of the life of the ocean surface and shallow waters, because it depends on the products of the surface and, in turn, contributes to the chemistry of surface waters, if somewhat distantly.

Possibly fortuitous examples of this interrelationship between living organisms are found in the many Archaean greenstone belts that contain stromatolites in close proximity to volcanic rocks. Some of the oldest probable stromatolites, in Barberton, seem to have actually grown on lavas. In many Archaean volcanic rocks there are hydrothermally produced veins, full of sulphides such as pyrite and often containing what is loosely described as 'graphite', or carbon-rich compounds. It is possible that much of this carbon is ultimately of organic origin, though it has since been metamorphosed. Near the volcanoes, in shallow seas and coastal lagoons, stromatolites must have been growing. We can imagine the seas of Barberton, Belingwe or Steep Rock. In part of each region, active volcanoes erupted; around them communities of bacterial life flourished in a hydrothermal setting, while in the water above and on the coast the cyanobacterial communities were producing oxygen.

This is simply an illustration – the planet as a whole is the ecosystem, not one locality, especially as water and gases mix rapidly and local anomalies in the seas (for instance, in the oxygen content of the water) would be shortlived. It is possible to imagine a complex planetary ecology even in the earliest days of the geological record. More than that, this complex planetary ecology was not some strange alien world, but the commonplace world of today. The same players act today as then, even if they have assumed modern dress. The chloroplast in the cell of the modern plant is simply a cyanobacterium in another guise. The mitochondria in our own cells are symbiotic descendants of bacteria. Or imagine the mammalian digestive tract: the methane-generating bacteria (methanogens) are at work as they have been for aeons. Plants and animals continue the old roles of the bacteria, and the system has been conservative over the aeons. Bacteria have simply found an efficient way of expanding their influence, by inhabiting us, our relatives and friends (e.g. cows). The important chemical reactions that fuel the biosphere were first used by the bacteria: we still depend on these reactions and the bacterial ways of carrying them out.

THE DOINGS OF ARCHAEAN LIFE: THE CREATION OF THE BIOSPHERE

If a complex web of life was created on the surface of the Earth four billion years ago, what were the results by the end of the Archaean? Was life simply a superficial veneer on the surface of the planet, or did it play some deeper role in the evolution of the Earth? The answer is that life is much more than an accidental smear of lipstick on the face of the Earth. Life has imposed a deep, pervasive pattern on the planet, shaping it as our home while simultaneously adapting to that home. Life is the architect of the planet's surface. It has accomplished this because it makes waste.

Very early in the planet's history some of the first organisms must have converted sugar to energy in the form of a nucleotide known as ATP (adenosine triphosphate), producing alcohols and acids. Fermenting processes can break down sugars, carbohydrates, amino acids, alcohols or acids, to give a few molecules of ATP and waste products such as carbon dioxide and ethanol, or lactic acid, or acetic acid and ethanol. The waste products of one fermenting bacterium can be used as the food for another, so from this a food chain can be established. Eventually the end products, such as carbon dioxide, are emitted to the environment, for instance to the atmosphere.

In the face of life the standing resources of a planet are small. Life reproduces; it would have rapidly used up all the available inorganically made resources. The food would soon have run out. A pair of rats placed in a corn store will eat and reproduce, until the store is filled with rats and the corn is all eaten. The rats will then eat each other; then they will die out. The first living organisms must have faced this crisis, but they did not die. Like the rats, they ate each other, fermenting dead bodies of other organisms (and living bodies too, on occasion), but the store was not emptied. Life found ways of gathering energy and discovered new supplies with long-term prospects.

The largest and most reliable source of energy is the Sun. Volcanoes come and go, but the Sun is always there. **Photosynthesis**, the transformation of energy from sunlight into the chemical energy of food molecules, is the basis of virtually all of the activities of life; it has fed us for billions of years. Even today, directly or indirectly, it provides most of the needs of humanity. It has also allowed life to manage the atmosphere.

Photosynthesis began in anaerobic bacteria (bacteria that live in environments without free oxygen), probably about four billion years ago, or even before although, as mentioned, the firmer evidence of bacteria only dates from around 3.5 billion years ago. The essential feature of photosynthesis is that with the aid of light energy, carbon dioxide (CO_2) is incorporated from air into organic compounds needed by the photosynthesizing organism. The energy captured from light is held as chemical energy, or food. This food can then be moved around by using the chemical known as ATP.

In order to carry out photosynthesis, a source of electrons is needed. These come from compounds containing hydrogen, such as hydrogen sulphide or water. In the first organisms to use photosynthesis, hydrogen sulphide may have been the hydrogen donor. Hydrogen sulphide is common in and around volcanoes, and perhaps the bacteria living in shallow water around active volcanoes may have been the first to use photosynthesis, gathering energy not from the volcanoes, as at first, but from the Sun.

Modern plants, algae and cyanobacteria do it differently. Instead of hydrogen sulphide, which is now relatively rare over most of the Earth's surface, they use water. They take water and carbon dioxide, and sunlight, and produce carbohydrate and oxygen. Somehow the ancestral cyanobacteria developed a second reaction centre for photosynthesis which could use high-energy light to split water molecules into hydrogen and oxygen. The hydrogen can then be used together with carbon dioxide from the air to make organic chemicals such as sugars. The oxygen is lost and emitted. Water and carbon dioxide are common; sunlight is a free and constant gift. An ecosystem can be set up, some organisms emitting carbon dioxide, others consuming it and emitting oxygen. Once this can be done, life can spread; it does not need to cluster around volcanoes or sources of hydrogen sulphide. The first use of this method of photosynthesis lies deep in the past, certainly before 3.6 billion years ago, possibly even before four billion years ago.

This gives an imaginary chain of events, one possible scenario for the early history of life (but, caution, there are many other possible scenarios, and there is no scientific agreement about which is correct). In this particular scenario, the first life began in a volcanic region around a subaerial hydrothermal system, using the available chemicals and spreading into shallow water from volcano to volcano, and to the mid-ocean ridges. But the available chemicals were scarce, and some living organisms evolved that were able to recycle the dead bodies of others. Other organisms, living in shallow water, were able to use hydrogen sulphide from the volcanoes and sunlight, to photosynthesize and so to increase the quantity of energy available dramatically. Eventually, a bacterium evolved that was able to use water and carbon dioxide. Both H_2O and CO_2 were abundant across the planet, and the biosphere could spread out from the original volcanic homes across the face of the Earth.

This, it must be stressed, is only one scenario. There are many other possible pathways that life could have taken, depending on which energy-capturing process evolved first.

The Sun is a free lunch. But there is also a catch: environmental degradation. Imagine the primitive ecosystem around a volcano: bacteria using hydrogen sulphide (H_2S). The sulphur in the H_2S is itself cycled by a complex set of processes. Some of the sulphur may be primeval, but much is reworked by a set of reactions in the hydrothermal system. In the hydrothermal system, seawater enters the hot rock. Sulphate in the seawater reacts with iron in the rock to produce H_2S and iron oxides, dissolving sulphide minerals in the rock. The hot water then rises back towards the ocean floor, carrying with it a heavy burden of sulphide. At the sea floor the H_2S is then taken up by bacteria that also take up oxygen dissolved in the seawater, combining them to make sulphate. Elsewhere, in shallow water in sunlight, the H_2S is used photosynthetically with CO_2, and sulphur is excreted.

There is a problem of balance here. If the ocean is poor in oxygen, sulphate is not produced, is not available to enter the rock, and less H_2S will then be available in the hydrothermal fluid. Perhaps the early ocean was oxidized by oxygen left over by the loss of hydrogen to space from the top of the early atmosphere. Or perhaps the system worked at a low level of productivity, using whatever H_2S was available in an oxygen-poor setting. But there is a need for an oxygen supply if the system is to work well.

The problem of balance would have become complex when photosynthesis involving water first began. In this process, water and carbon dioxide, with sunlight, produce sugar and oxygen. The cyanobacteria use light, CO_2 and water to produce organic food chemicals such as sugars. They excrete oxygen. Most larger organisms on the modern Earth depend on the results of this type of photosynthesis, whether the organisms are producing-photosynthesizers (plants) or consumers breathing the oxygen in the atmosphere (animals).

On the early Earth, the oxygen produced by photosynthesis would have been taken up, via seawater, by the iron and sulphur in the rock through which the hydrothermal systems operated, and also by iron and sulphur in material eroded from any land areas that existed. Here is a cycle that can operate steadily. Oxygen and sulphates entering hydrothermal systems produce H_2S; H_2S is used by bacteria both in photosynthesis, capturing CO_2, and with oxygen to make sulphate; other bacteria photosynthesize, using H_2O and CO_2, liberating oxygen that finds its way back to the hydrothermal systems. The populations of all the types of bacteria are limited by the availability of the resources, which are themselves controlled by the activities of other bacteria by the input of sunlight and the vigour of the hydrothermal system. In all, this is a simple ecosystem.

But there is a further problem. CO_2 is consumed by both the bacteria that photosynthesize using H_2S and those that use water. So CO_2 would have been eaten up and removed from the air. Yet the greenhouse effect of CO_2 was essential to keep the planet warm, in the face of the faint young Sun. If too much CO_2 were eaten, the planet would cool, the seas would freeze and life would be reduced to a precarious existence in a few hotsprings and volcanoes, until more CO_2 degassed from the mantle via the volcanoes. This new CO_2 would then be eaten up by life, and so on: a biological catastrophe that could have led to the extinction of life.

In his book *The ages of Gaia*, J.E. Lovelock has suggested that the actions of early methane-generating bacteria (methanogens) helped to ensure that the catastrophe did not occur. The methanogens are archaebacterial scavengers. They would have lived in decomposing organic matter – bacterial bodies – converting carbon from the organic matter to methane, which would return to the air. Lovelock pointed out that the methane released by the methanogens could have played a critical role in protecting the Archaean biosphere. Methane is a powerful greenhouse gas, and so it would have helped to keep the planet warm. Moreover, ultraviolet light acting on the methane in the air could have produced an upper atmospheric smog, which would have protected the living organisms below from the solar ultraviolet irradiation. Perhaps such a smog layer was the Archaean equivalent of the ozone layer in the stratosphere of the modern Earth. Lovelock's hypothesis is speculative, but quite possible.

How much oxygen was there in the system? We do not know. The only evidence is the presence of sulphate deposits in the Pilbara rocks, which implies that at least some oxygen was chemically available in the seawater, though free oxygen was not necessarily present in the air. On the other hand, if the young

Earth ever passed through a phase in which it approached a runaway or moist greenhouse catastrophe, much free oxygen would have been left behind in the atmosphere as hydrogen from water was lost to space. Perhaps there was a delicate balance, with oxygen being a minor component in the air (possibly 0.2–0.4%) – enough oxygen in seawater to make sulphate, but not enough to build up as a substantial constituent of the air.

Now, the release of oxygen by photosynthesis is excellent, as far as human beings are concerned,but too much oxygen could have poisoned the early bacteria. We do not know how much oxygen existed in the air and sea during the early history of the biosphere. As mentioned above, the oxidation of iron could have controlled oxygen levels, but too much free oxygen would have made it impossible for many bacteria to live on the very surface of the planet. If so, how could anaerobic bacteria manage to live, and how do they survive today? The answer is that, even today with abundant oxygen in the air, much of the biosphere operates in the absence of oxygen. It is easy enough to show this, just by pushing a stick into the mud at the bottom of a shallow stagnant pond. The bubbles that rise up will most probably be rich in methane ('swamp gas'), made in an environment in which oxygen is absent. If the 'biosphere' is the thin layer on the Earth's surface which sustains life, then much of that layer, especially just below the sea floor, is poor in oxygen. This must have been the case throughout the Earth's history.

We do not know what sort of energy supply – volcanic or sunlight – came first, and if we did we might come close to understanding where life began, whether in a hydrothermal system, a shallow-water mud or a small surface pool. But once established, the living community must soon have diversified its use and its methods of gathering energy. At a guess, fermentation may have come early, exploiting the waste and debris of the first chain, and creating a new subchain of its own.

At first glance, nitrogen is a rather inert gas, simply stored in the air. This is true of N_2, that makes up the bulk of the air. In other compounds, however, as in NO_2 for example, the nitrogen is not inert and is an essential component of life. Fixing nitrogen, changing it into a biologically useful form, is one of the most important processes in the biosphere. Lightning fixes quantities of nitrogen: on the modern Earth, lightning strikes most over the rainforests, especially in the East Indies, over Borneo, and also over the Amazon. When the land was uninhabited by life, in the early history of the Earth, this source may have been smaller than today. But lightning is a minor source of useful nitrogen even today. Most biological nitrogen available to the natural land biosphere is fixed by the bacteria that exist symbiotically in the soil with plants. The first nitrogen-fixing bacteria probably evolved early in the history of life, long before the plants, though today nitrogen-fixing bacteria co-operate with plants. For instance, peas and beans can fix nitrogen because they host symbiotic bacteria in root nodules. When the pea and bean plants die in the classic English vegetable garden, they are turned into compost and their nitrogen is recycled into the soil.

Here we have another potential crisis: why was the supply of biologically usable nitrogen not exhausted? In the beginning of the ecosystem, was the total biomass of living organisms once limited by the availability of useful nitrogen, which was supplied inorganically in small quantities by lightning that fixed it from the air? Later, once nitrogen-fixing bacteria had evolved, what would happen when all the nitrogen in the air was used up? Fortunately, the cycle has returns too. Some organisms emit nitrogen. These denitrifying bacteria take fixed nitrogen compounds and produce from them nitrogen. This process occurs today in quiet, stratified ocean water and in soils (e.g. in rice paddy-fields). Earth's atmosphere also has a small but very important amount of biologically made ammonia. This would not exist inorganically in the air because it is either rapidly rained out or it is dissociated by light in a few years. To life, ammonia is helpful, as it controls the acidity of rain – without ammonia, rainfall would be as acid as vinegar. All these processes add up today to a complex biological cycle of nitrogen, cycling the gas back and forth between the air and living organisms.

Nitrogen-fixing bacteria seem to have evolved very early on in the history of life. Fixing nitrogen takes a large amount of energy, and if these bacteria had not evolved, life would have been limited by nitrogen shortage. Instead, a nitrogen cycle was set up, some bacteria releasing nitrogen to the air, others recapturing it for use by life.

BUILDING AN ECOSYSTEM

We can now imagine a more complex early world of bacteria, some living in hydrothermal systems and oxidizing sulphide, some fermenting the waste, others in shallow water, possibly in subtidal mud-flats, removing hydrogen sulphide, and – in a wilder imagination – the whole system perhaps helped along in its early days by a slowly declining small loss of hydrogen to space and a consequent small supply of oxygen. Simultaneously, the living community was improving its ability to manage the nitrogen balance of the biosphere. But this is not a secure system – it would eventually have died. It has no way, other than good luck, of surviving the changes as the Solar System aged, the Sun brightened and the planet recycled its surface chemistry, and erupted CO_2. Eventually the Earth would probably have gone the way of Venus. That is, if Earth had been populated by organisms which did not evolve. But natural selection forces life to change.

Life attains a moving equilibrium. There is no equilibrium in time, no stasis. Evolution is an upward spiral. Over the medium term, populations balance, or at least fluctuate around a norm. In a large African game park today, all other things being equal, the number of lions depends on the number of prey, such as antelopes, and the number of antelopes depends on the number of lions. If too many lions are born, some will die of hunger; if for some reason extra antelope are added, the well-fed lions will multiply, gobble them up and, eventually, die off until the equilibrium is restored. Over the longer term, equilibrium is disrupted by evolution: faster antelope survive and reproduce, more skilful lions catch them. The competition drives natural selection, and new species appear.

All this, of course, is much too simplistic; perturbation is infinite, but it allows us to make a simple analysis of the early Archaean planet. For instance, once evolution had produced both nitrogen-fixing and denitrifying bacteria, the level of nitrogen in the atmosphere would be set when a balance was struck between nitrogen production, by bacteria and lightning, and nitrogen consumption, allowing both consuming and producing bacteria to maximize their numbers. Even more important, carbon dioxide would be managed. With the evolution of oxygen-releasing photosynthesizers, CO_2 control became possible, because the cyanobacteria would spread explosively into every available well lit niche, fixing carbon out of the atmosphere, and allowing other bacteria, such as those in hydrothermal systems, also to live and to fix carbon, balancing the emission of CO_2 from organisms and volcanoes. Eventually – and perhaps rather soon in geological time – a rough equilibrium would have been struck, limited by the supply of other nutrients such as phosphorus.

The balance is not perfect. Volcanoes erupt, and place dust, sulphur gases, and CO_2 into the air, at irregular intervals. The eruptions can disrupt the climate for long periods, especially if sulphur gases enter the stratosphere (for example, the eruption of El Chichon may have contributed to the massive changes in the world's climate in 1982–3). The CO_2 acts on a longer time span as a greenhouse gas warming the planet. If the CO_2 content of the air increases, organisms grow faster and multiply in the new conditions: the biosphere reacts by fixing the carbon and removing it. Or, if the biomass of life grows too much, it takes away the protective greenhouse of CO_2 around it: the world cools and growth slows until decay or volcanic gases restore the earlier equilibrium. Reality, of course, is today much more complex, with innumerable subcycles, but the basic feedback loops can be imagined.

Even from the early Archaean, we can imagine an atmosphere that was coming increasingly under the influence of life. The carbon isotopes in old sediments tell us that the carbon cycle was biologically controlled 3.5 billion years ago. If the fragmentary inference from the carbon isotopes of the very old sedimentary rocks from Isua in Greenland is to be believed, there is evidence that this biological control of the carbon cycle may even be as old as 3.8 billion years. If so, we should also assume that by this time life controlled the composition of the atmosphere and that the control has been maintained ever since.

Control of CO_2 (and CH_4) means that the surface temperature of the planet is controlled.

The consequence of temperature control, if it occurred, is that liquid water remained stable on the surface, despite the steady warming of the Sun over the aeons. We must assume that the inorganic, pre-biological Earth was, by chance, at the right distance from the Sun and had the right atmosphere to allow free water to exist for at least several hundred million years – enough time for life to get started. Lovelock's Gaia hypothesis is that, ever since then, life has maintained a nearly steady surface temperature by managing the levels of CO_2 and other greenhouse gases such as CH_4. The alternative hypothesis is that the more than 4 billion year history of liquid water on Earth has been produced by blind inorganic chance. As yet, we do not know which hypothesis is true or, if it is accepted that biological controls do control the environment *now*, when that control began.

It is possible that the atmosphere has been biologically controlled ever since the onset of life, though obviously the extent and sophistication of this control must have slowly increased over time. Initially, life could have exerted only a long-term influence on the atmosphere: today it is possible that virtually the whole composition of the atmosphere, except for the rare gases, is probably under biological direction. All this is, of course, scientific imagination. We cannot yet *prove* that the Archaean atmosphere was biologically governed. Perhaps life has exerted environmental control throughout the past four billion years, since the time before the beginning of the geological record, preceded by a 'brief' few hundred million years during which the oceans existed in an essentially inorganic atmosphere, near to chemical equilibrium and controlled by the chemical reactions (e.g. calcite deposition) taking place at hydrothermal systems. Incidentally, it is worth remembering that even the brief period before four billion years ago was a time as long as the whole history of the fossil record of the hard parts of organisms, from the Cambrian to the present.

The deduction follows that, if life did manage the atmosphere, it must also have helped to shape the evolution of the surface and interior of the planet. The effects of life even reach to the core of the world. This assertion is another way of stating the claim, made earlier, that life is the architect of the planet.

We can explain this by comparing Earth with Venus. On modern Venus the surface temperature is 500°C (900°F). If all the CO_2 now locked up in limestones on Earth had been allowed instead to accumulate in our atmosphere, the temperature here too would be something out of the Inferno. On Earth we have an assortment of rocks that probably do not exist on Venus. These rocks, which are formed in cool, high-pressure conditions, include the rocks known as blueschists that are common in Western China, California and parts of the Alps. Temperatures on Venus are too high for these rocks. Eclogites, which are dense rocks which help in the driving of plate tectonics, are also unlikely to exist on Venus. Sadly, Venus may have few diamonds to adorn her – they probably can't exist there. More generally, the low-temperature hydrous granitic and metamorphic rocks that make up the bulk of the Earth's continents also cannot exist on Venus.

All this is the direct consequence of the surface temperature. The lack of water imposes another, greater constraint on Venus. Remember the hydrothermal system at the Earth's mid-ocean ridges, the radiator on the front of the Rolls-Royce that is the Earth? Remove the radiator and the Rolls-Royce not only ceases to be pretty, but seizes up. Venus, the Volkswagen of the planets, is air-cooled instead, and just as an old Volkswagen's air-cooled engine is designed differently from that of a Rolls-Royce, so must the planetary heat engine of Venus be different from the Earth's. The cooling on Venus is far less immediate and far less effective, and Venus has probably adopted a different strategy of operating its surface. Heat, in lavas, may erupt in many regions. In contrast, the Earth's plate system is highly organized, with a significant part of the heat loss from the mantle directed along the lines of the mid-ocean ridges. Furthermore, heat loss from the interior to a surface at 500°C (900°F) is very different from heat loss to a surface at 15–25°C (50–80°F) – Venus probably has a hotter deep interior than Earth (which of course one would expect from the hot surface). Even the state of the core of the Earth, the inner part of which is solid, the outer part liquid, is critically dependent on the temperature and pressure regime in the planet. The core of Venus may have a different structure from that of the Earth.

Water is not just the coolant in the radiator: it is also the flux, or agent that helps melting, allowing extensive partial melting to occur at relatively low temperatures in the mantle above subduction zones, and promotes the formation of granites. It has been suggested that without water we could not have granites, and without granites we could not have continents. We do not know if there are continents on Venus. Perhaps there are continents on Venus made of a silica-rich melt called tonalite (a granitic type that exists on Earth and which needs little water in the melting process), but the broad chemical composition of the surface of Venus must be different from the Earth.

On Venus, erosion is by wind, not water. There is no clear division in the altitudes of the surface (what is known as the hypsographic curve) between low-lying areas and high areas. The surface of Venus grades from very high terrain, such as the immense Maxwell mountains, to very low terrain. In contrast, Earth has distinct continents set above sea level, and broad oceanic plains that are 3–5 km (2–3 miles) below the sea surface. Some of the sharpness of the division of altitude on Earth is simply a product of water erosion – this is what gives the continents their nearly uniform thickness and their great plains – but there is a well defined division between regions of silica- and alumina-rich granitic crust that make up the continental areas and regions of magnesium and iron silicate crust, the oceanic floors. Whatever system dissipates heat on Venus, it is not the same as on the Earth. And there are no water-lain sedimentary rocks on Venus. Venus is not a primitive Earth: it is an evolved planet, but it has evolved in its own way.

The whole style of the two planets is different. And yet they came into existence so similarly, with much the same diameter (Venus is a little smaller) and probably with oceans of water under a cloudy sky, rich in CO_2. Venus is much closer to the Sun, which accounts in part for the present difference. But why did the Earth not follow the path of Venus, somewhat later? That the Earth is so different in surface, in its geological style and perhaps even at its core, may be a consequence – at least in part – of the existence of life and its management of the greenhouse. We, living beings, may not be just a surface accident. We may have helped to design and create our home.

FURTHER READING

Cloud, P. 1988. *Oasis in space*. New York: W.W. Norton.
Lovelock, J.E. 1988. *The ages of Gaia*. New York: W.W. Norton.

6
The surface of the Proterozoic Earth

THE MIDDLE AGE

In the imagination of the geologist, the 'Proterozoic' – the two billion years between 2500 million years ago and about 570 million years ago – is the time of great but shadowy events. The dark beginnings were closing: the world was starting to take its modern shape. In the late Archaean, we see the biosphere in its medieval morning, dispelling the Dark Ages. Then, at some time in the Proterozoic the bacteria give rise to the eukaryotes, with their cellular nucleus and sexual reproduction: courtship begins, and jousting between males – the High Middle Ages of our planet. At the end of the Proterozoic come hard skeletons and high-tech predators with good eating equipment: strong animals with manifest destiny. The modern 'Phanerozoic' world has begun. Such is the teaching of the geological textbooks. It may be true, or it may not. Perhaps evolution is not inevitably bound to follow a single path – there may have been many other pathways that, by accident, were not chosen. Had one of these alternative paths been followed, life today might have been very different. The history of life is rich in dividing points where chance or the hand of the deity prevailed. Perhaps we owe more than we can imagine to our bacterial predecessors during the Dark Ages.

What were the sagas of this middle age of the Earth? Geologists see the Proterozoic as a time when the great continental aggregations became established and the planet slowly took on a modern aspect. Life was there, and its deeds were great. Before investigating the history of that life, though, we need to build up a picture of the physical nature of the Earth's surface – the habitat of life.

BUILDING THE CONTINENTS

We do not know how large the continents were in the Archaean. There are many theoretical models of continental growth, some of which suggest that huge continents have existed since the dawn of geological time, and others that most of the continental growth took place as late as the Proterozoic. Geologists are divided over the validity of these models: the most popular hypothesis is that much of the continental growth took place towards the end of the Archaean and in the early Proterozoic. But science is not democratic: the less popular models, which suggest that large continents existed at the beginning of the Archaean, could instead be true.

We do know some facts though. The continents as they exist today, and as they are represented in the sediments eroded off them, are on average very roughly two billion years old. This means that there must have been much recycling of the continents into the mantle if the continents first appeared very early in the planet's history, say, four billion years ago. Alternatively, if most continental rock was freshly generated, say, two to three billion years ago, less recycling would have taken place. Even today, we know continental processing is taking place: for instance, India is disappearing under Tibet; elsewhere, sedimentary material derived from erosion of continent is being carried downwards in subduction zones and taken back into the mantle by many oceanic trenches; the volcanoes ringing the Pacific are bringing material back up to the surface of the continents. If we look at the continents we see that they are a patchwork aggregate of material of different ages. North America is a fascinating example of this, and worth considering in detail.

The construction of North America

North America is composed of a variety of Archaean fragments, held together by liberal amounts of Proterozoic cement (as much cement as bricks) and variously covered with younger material. The Archaean bricks include material up to four billion years old in the Canadian north-west, and up to 3.8 billion years old on the Labrador coast, but more central to the present continent is the nucleus terrain that lies in what is now northwestern Ontario, Canada, between Thunder Bay and Winnipeg. The core of this region is around three billion years old. About 2.6 billion years ago the huge Superior region was mostly in being, as was the separate Slave region in northern Canada. Today, in these areas we can map out large late Archaean belts of volcanic and sedimentary rock, separated by belts of granites. In the Kapuskasing area in the middle of the Superior province we can see deeper into the crust, where a massive fault and subsequent erosion have exposed material that may have formed at depths of up to 30 km (20 miles), in the crust. This material from the deep crust is a dry and once-hot metamorphic rock known as granulite, similar

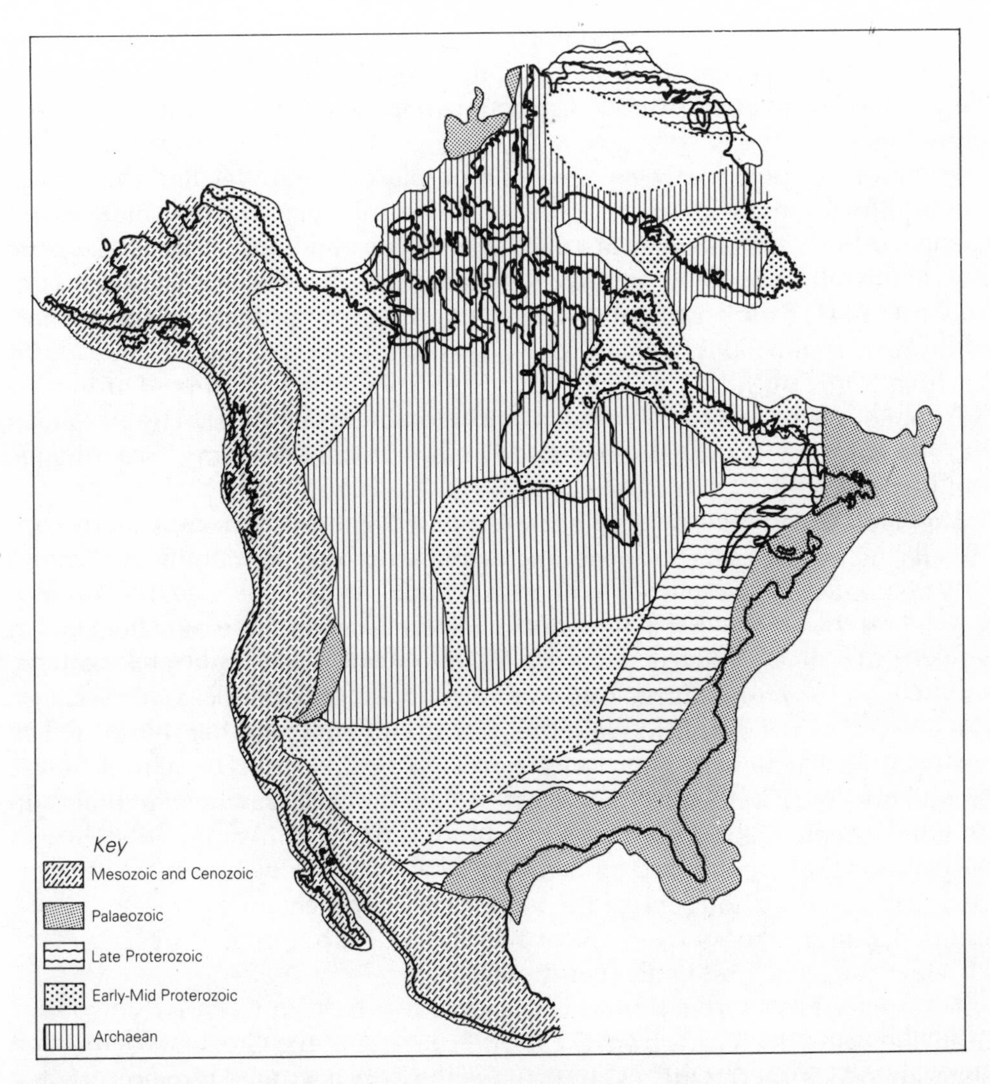

Figure 6.1 The accretion of North America. Map shows the time when the major parts of North America were welded on to the Archaean nucleus (with thanks to H. Williams).

to material exposed today in young belts such as the Alps and Himalayas.

These are the early continental fragments: the Slave and Superior regions, and the fragment of Labrador (then attached to what is now Greenland and northern Scotland). Where on the globe they were, we do not know. They may have been isolated microcontinents, or more probably they were aggregated to other fragments, such as the centre of Zimbabwe, part of South Africa, or Western Australia, or fragments now destroyed. Continents move, drifting slowly across the surface of the globe at the rate of a few centimetres per year as the tectonic plates are created and destroyed. If we are lucky, measurements of old magnetic directions can give us the latitude in which rocks formed, but the problems of dating later deformation, and alteration are so immense that it is not yet possible to attempt a reconstruction of the continents at the close of the Archaean.

In the Proterozoic, however, events took place that aggregated these fragments into the nucleus of a continent which included North America and Europe. About 1.8 billion years ago, a mountain-building event, called the Trans-Hudson Orogen, cemented together the Archaean bricks into a continent. This 'Hudsonian' event produced mountains comparable to the Alpine–Himalayan chain today. Their eroded remnants can be traced across North America from South Dakota across Hudson Bay and into Europe. During the Hudsonian event, several Archaean microcontinents collided. The Archaean fragments included the Superior and Slave regions, Wyoming, Scandinavia and part of Russia.

The Hudsonian collision is best preserved in northern Saskatchewan, Canada, where the edges of the microcontinents can be identified. Around 1900 million years ago two separate microcontinents existed, one the Superior region and the other what is now northwestern Canada. Between them lay an ocean, and a major volcanic arc developed in what is now central Saskatchewan. Gigantic granitic intrusions were emplaced, and the ocean closed in a massive collisional event, to weld the microcontinents together into one. The centre of Saskatchewan must once have looked like the Himalayas today, though now most of the province is now so flat that a Saskatchewan Mountain Club has adopted as its motto 'No mountain is too low'. Among the relicts of the collision that are exposed today in the forests of Saskatchewan are an enormous granitic intrusion (one of the largest on Earth), chains of volcanic rocks, and metamorphic rocks which have been brought up from depths as great as 25–30 km (approx. 15–20 miles) or more.

A younger Proterozoic terrain is preserved in Eastern Canada. This is the Grenville Belt, which runs from the northern shore of the Great Lakes through Ontario and southern Quebec. It includes the deeply eroded remains of what was probably another massive continental collision. Perhaps this entire belt of rocks was once the roof of the world, like Tibet today. Along the belt, erosion has stripped off the upper crust of the Earth and exposed the deeper layer of the continent. Hundreds of millions of years from now, this will probably also happen to Tibet. This is because the crust acts rather as blocks of wood floating in water. As erosion strips off a layer of crust from the top, the block rises, and more is eroded, and so on until balance is attained. To erode a mountain like Everest (8 km; 5 miles high) to flatness, about 30–35 km (approx. 20 miles) of crust must be removed before the block reaches equilibrium. Eventually, the mountains are reduced: the cities of eastern Canada now stand on the remnants of the mountains that were built up and then eroded on the flanks of the old nucleus of North America.

During the Proterozoic, the forerunners of the modern continental masses were constructed, though we do not yet know to what extent this construction used the fragments of yet older continents, or whether much of the continental material was newly separated out of the Earth's mantle. By the Late Proterozoic most of the granitic rafts that we call continents were in existence. Since then the rafts have been broken into fragments and then re-aggregated in different ways by continental movement, or drift. Some special areas have remained very little altered since the Archaean. If we had dropped a soft-drink can on the surface of the Steep Rock or Belingwe beach nearly three billion years ago we might hope to find its remains eroded out today.

THE SEDIMENTARY BASINS OF THE LATE ARCHAEAN AND PROTEROZOIC

Sedimentary basins are thick accumulations of sedimentary rock, preserved when they are laid down on continental crust. Most Archaean sedimentary rocks are now found in a highly deformed state. In Zimbabwe, field geologists used to joke that Archaean sediments were easy to map because they had all been tilted around so that they dipped vertically. This is true in many areas of the world. Most Archaean sedimentary rocks have been folded, heated or compressed. In southern Africa and Australia, however, there are a few late Archaean successions of rock that are not folded and, for the most part, are still recognizable as sedimentary rocks, in a state close to that in which they were laid down. In the Proterozoic, in contrast, little-deformed successions are common and they occur across the world; many large sedimentary accumulations or basins have been preserved. More recently, in the Phanerozoic, the formation of basins has continued. In North America, for instance, the Michigan Basin, the Williston Basin, and the Illinois Basin have developed and in Europe there is the North Sea Basin (which is still collecting sediment).

One of the most famous of the old basins is the Witwatersrand Basin of South Africa, which is of late Archaean age. The rocks were probably laid down in a large inland sea, about the size of Lake Victoria, into which poured the detritus eroded off the surrounding countryside. The centre of the sea was subsiding, and the added load from the sediment amplified the subsidence. Eventually sediments such as sands and gravels several kilometres thick were laid down in the basin. The detritus from the Archaean countryside around was rich in gold, and the heavy gold particles dropped out as the rivers that carried them entered the sea. Biological activity may have played an important role in this process. Two and a half billion years later these sedimentary deposits became the source of much of the world's gold. The diamonds carried by the rivers have already been mentioned. They too were dropped out, into the sediment.

In central South Africa, the Earth's crust and lithosphere was thick, cool and strong by the end of the Archaean. Beneath the Witwatersrand, the temperatures in the lower parts of the crust and in the upper mantle were fairly cool. Cool rocks are strong rocks, and these strong rocks were able to bear the burden of the thick sedimentary basin on top of them. Occasionally, eruptions must have passed through the lithosphere to carry diamonds up narrow pipes to the surface, but in general the lithosphere here was thick, stable, and undisturbed.

In Western Australia a similar succession of little-deformed sedimentary rocks lies on the old Archaean crust. This is the Hamersley succession, which is about 2500 million years old and formed around the time of the boundary between the Archaean and Proterozoic Eons. The Hamersley rocks are famous for being so little deformed that one can trace individual bands of rock across tens or even hundreds of kilometres of outcrop. The Hamersley Basin is one of the main sources of iron ore for Japan, and millions of tons are shipped out each year: many of us have parked on our drive a ton or so of Hamersley iron, originally laid down two and a half billion years ago.

Figure 6.2 Ironstones of the Hamersley Basin, Western Australia. For scale, see trees.

The iron ore occurs in banded ironstone, a sedimentary rock that forms a significant proportion of the layers in the Hamersley Basin. The ironstones occur in extensive bands, which may cover thousands of square kilometres. The bands seem to have been laid down in very quiet water, and they may have been formed as a result of the activity of bacteria that precipitated iron. Ironstones of various types have formed throughout geological time, but the deposition of banded ironstones over tens of thousands of square kilometres is a distinctive feature of Proterozoic geology. Interestingly, Proterozoic ironstone is so heavy that, as it filled a sedimentary basin, the basin would have continued to subside under its own weight and for no other reason, because the sediment added on top was heavier than the mantle displaced beneath.

The Proterozoic sedimentary basins not only have ironstones; some are also marked by thick successions of carbonate rock. In the Archaean greenstone belts, some minor limestone occurs, including the stromatolitic rocks in the Steep Rock and Belingwe Belts, as well as in the late Archaean rocks of Western Australia, such as the Fortescue Basin in Western Australia, where there are well developed and well preserved stromatolites. But for the most part, Archaean stromatolites are rare, as are Archaean limestones, many of which have furthermore been highly deformed by later events. In contrast, in the Proterozoic basins, little-deformed stromatolitic rocks are extremely widespread. Enormous extents of almost intact stromatolitic rock occur, laid down in the shallow and intertidal waters of vast shallow seas. These Proterozoic rocks today are mostly dolomite (magnesian carbonate), rather than limestone (calcium carbonate); no one knows why, although there are many theories.

Figure 6.3 Proterozoic stromatolites, near Cluff Lake, northern Saskatchewan, Canada. This inverted V structure is characteristic of many stromatolites.

The Proterozoic continents hosted many large inland seas, in which grew abundant bacterial colonies that were later preserved as stromatolites. There are few close modern analogues to these seas, except in warm bays around Western Australia and in the Bahamas. This is partly because modern bacterial and algal colonies are cropped by animal grazers. Bacteria and algae still grow abundantly, but most are eaten as fast as they grow. They have little chance to produce massive stromatolite structures, except in a few very protected settings such as salty lagoons where the grazers are inhibited.

Large continental sedimentary basins are not common today. This is partly because we live in a time of very unusual climate, with enormous amounts of water locked up in the ice caps. Today, after continental breakup and, more recently, the ice ages, the planet is not in erosional equilibrium, and most of the inland basins that should exist have dried up. A few places, such as Lake Chad and the Okavango swamp in Africa, the Aral Sea in Asia, Lake Eyre in Australia, and perhaps the North Sea, remind us of the huge continental inland seas of the past. Were sea level to rise a hundred metres or so, many of the great cities of Europe and sizeable chunks of most continents would be inundated, and deposition would begin again. If rainfall increased, the inland deltas of Africa and Australia would become great lakes occupied by complex communities of life. If the continents drifted together again and, over a few tens of millions of years, were eroded down towards sea level, large inland basins would reappear in places of rifting or where the lithosphere was subsiding.

We can imagine the Proterozoic basins, shallow inland seas fringed by stromatolites where light could reach the photosynthetic organisms, and the sites of ironstone deposition in quiet water. These basins must have lain on large continents, which for the most part were already ancient and eroded nearly to flatness (like modern Australia). The Earth is a dynamic place and many sedimentary basins are destroyed, uplifted and eroded by mountain-building events such as continental collision, yet some basins have survived little deformed through to our time, having quietly sat on their continental rafts over the billions of years. They have all been heated a little, of course, by burial under a blanket of younger sediment, or by conduction from nearby igneous events, but in some basins the old rocks are nearly intact. It is these rocks that give us some of the best evidence available for the history of life in the Proterozoic.

The sedimentary story of the Proterozoic basins is a quiet tale, of shallow waters and muds interspersed with clear lagoons, in which stromatolites grew. If we go on vacation to the Bahamas, we can also see quiet sunny lagoons and swim above algal mounds. But a hundred million years ago the Atlantic was opening and the Bahamas were not so peaceful. A hundred million years in the future and the Atlantic Ocean may be closing again. Those algal mounds growing today in the Bahamas will probably be folded, buried and pounded into the Florida coast in a range of volcanic mountains. Over time, a few sedimentary sequences survive, others are covered by volcanic rocks, and yet others are folded, eroded, destroyed.

PROTEROZOIC IGNEOUS ACTIVITY

Many of the late Archaean and Proterozoic basins have thick suites of lavas interbedded with the sediments. More-deformed Proterozoic rock suites often include enormous volumes of lava and igneous intrusions: the Earth was, and is, a hot and active planet. Today, volcanism is mostly restricted to mid-ocean ridges, to island arcs (such as the Aleutians in the Pacific), and to the edges of continents, above subduction zones. This type of volcanism is common in the Proterozoic record. From time to time, a different sort of volcanism occurs on a large scale. In the geologically recent past (the past 150 million years), outpourings of lava of a type known as flood basalts covered enormous areas of the southern continents. In southern Africa, for example, during the Mesozoic several million square kilometres of land were covered by lavas: similar areas were covered by the Deccan lavas of India, and by lava in South America and Antarctica. Less impressive but still very extensive flood basalts covered large areas of the Columbia River Basin in the USA and also regions around the North Atlantic, including parts of Scotland, during the Tertiary Period.

This process has occurred at intervals throughout geological history and is especially well recorded in the Proterozoic. Some spectacular Proterozoic examples of these lava outpourings form the volcanic layers along the north and west flanks of Lake Superior. One of the largest of all igneous bodies preserved on the continents is the Bushveld complex in South Africa, which is very roughly two billion years old. This body is several kilometres thick. It contains some of the world's most important ore bodies for chrome and platinum. Up to tens of thousands of cubic kilometres of magma were intruded into the Earth's crust in a complex series of events and then cooled and crystallized. Each time such an eruption occurs, gigantic amounts of CO_2, trace gases and eruptive debris are released into the environment, challenging the stability of the biosphere.

THE PROTEROZOIC WORLD

Can we attempt to reconstruct the surface of the Proterozoic world? What would some imaginary Proterozoic astronaut have seen from a spacecraft orbiting the planet?

The most obvious features would be the same as today. The Earth had large oceans. There is geological evidence to suggest that large continents were set into those oceans, occupying perhaps a quarter or a third of the surface of the planet. Continental drift took place then as now: at times, supercontinents were present, while at other times in the Proterozoic an array of smaller land masses was spread across the surface. A keen geological eye might have recognized individual fragments such as the Superior Province of the Canadian Shield, but only if our astronaut were aware that in the later Proterozoic the Superior region was part of a continent that extended from what is now Canada across Greenland to northern Scotland, Norway, Finland and parts of Russia. Elsewhere, another fragment comprised what is now South Africa, Zimbabwe, parts of Antarctica, South America, India and Western Australia. Over billions of years pieces of continental rafts have wandered like drunkards across the globe, at times grouping together into larger continents, at other times disaggregating by rifting apart into smaller land masses and microcontinental blocks and leaking volcanic magma between.

The seas were probably as deep as today, or even deeper, because some water may have disappeared down to the mantle via subduction zones since the end of the Proterozoic. Beneath the seas the Proterozoic oceanic crust was probably rather similar to modern oceanic floor, except that it may have been slightly richer in magnesium and perhaps somewhat thicker than today, possibly formed at rather hotter mid-ocean ridges. The mantle below the lithosphere was probably somewhat hotter than today, perhaps by 100°C or more.

Almost certainly the dominant global process shaping the continents and oceans was plate tectonics, as today, though in detail the workings of the plates may have been slightly different. Occasionally, continental collisions took place, on a scale comparable to the event that today is producing the Himalayas as India crashes into Asia. The shuffle of the continents is clumsy, and periodically the drunken dancers crash catastrophically into each other.

So far, the picture is not too dissimilar from the modern world. But there are differences, and they would be obvious to the imaginary astronaut. The surface of the planet would have looked quite alien, with no dark green forests or grassy plains. Erosion would have been unchecked by plants as water rushed to the sea. Perhaps the surface was mostly a desert crust, inhabited in the early Proterozoic only by bacteria. The seas would have teemed with bacteria, not fish. The climate might have been very different. At times, ice-caps existed as today (there is good evidence for glacial periods in the Proterozoic), and generally, the climate was more or less within today's modern temperature limits or warmer, but the driving forces behind the climate would have been different. There was no Amazon forest to host thunderclouds that recycle enormous volumes of water between the atmosphere and surface, and no cover of leaves to ameliorate the extremes of the land surface and annually to manage CO_2. Far less heat would have been transferred from the Equator to the poles, implying a hotter Proterozoic equator, and colder poles. The composition of the atmosphere would have reflected the impact of life in the water: air was chemically processed by the life in the seas, as gases were exchanged between water and atmosphere.

FURTHER READING

Windley, B.F. 1984. *The evolving continents*, 2nd edn. New York: J. Wiley.

7
The birth of the eukaryotes

PROTEROZOIC LIFE

At the end of the Archaean our planet was dominated by bacteria. It still is, but today the bacteria are no longer in the foreground. Instead, they form the essential background to existence. The more obvious players on the stage today are the multi-celled plants and animals that have occupied the seas and colonized the continents, to form a complex, planet-wide, interdependent network of life. How did the change come about?

The descent of life: a summary so far

To recapitulate, the molecular record tells us that there are probably three ancient dynasties of organisms: the eubacteria, the archaebacteria and the eukaryotes. The two bacterial dynasties seem to be of the deepest antiquity, and their division goes back close to the dawn of life. The third dynasty, the eukaryotes, is more complex. They seem to have arisen by a symbiotic combination, or series of combinations, between bacterial ancestors from both sides of the divide. When this took place we do not know, but if molecular changes in the genetic message of organisms occurred at a roughly steady rate, the likelihood is that the eukaryotes originated a very long time ago indeed, probably in the mid-Proterozoic, but just possibly as early as the Archaean.

By the beginning of the Proterozoic, the living fabric of the planet was well established, with a bacterial economy of intricately interlinked organisms, each depending on the activities of others. Some bacteria used photosynthesis, to capture energy from sunlight. Others recycled the energy, for instance in fermentation processes, starting with sugars or carbohydrates and breaking them down. Yet other bacteria took different waste products and broke them down, deriving more energy from the process. Margulis and Sagan, in their book *Microcosmos*, which is a fascinating examination of the world from the bacterial point of view, liken the process to the activity of dung beetles today, which recycle the excrement of large animals. The analogy is excellent in Africa (where large mammals abound): innumerable dung beetles rapidly attack any dung and soon break it up. There is much benefit from this process, because the dung is carried around to help manure the plants in a relatively wide area and, further, the dung is quickly broken up and so does not last long – the habitat for breeding flies is limited. By stimulating plant growth and controlling disease-carrying flies, the beetles help their benefactors, the large mammals. In contrast, in Australia the native beetles are not properly capable of dealing with cow-pats, produced by an alien animal. Australia is now a continent infested with flies, something that comes as a great surprise to those used to working in the African bush where flies are typically much less common. Dung beetles are like sewers – they are modestly essential to a complex ecology. The same is true of the bacterial chain that breaks down waste. There is an old scatological tale of a Greek hero who flew to heaven on a dung beetle fuelled by his own dung. Life is like that, attaining the heights by recycling its own ordure.

By the end of the Archaean, our bacterial ancestors and cousins had created a diverse world economy in which some organisms used photosynthesis to take CO_2 and produce organic compounds and oxygen, while other organisms fed upon those compounds, all in an intricate and self-managing ecological web. Today the web is the same: although many of the chemical reactions carried out by the bacteria may be hidden in plants and animals, we have not changed the fundamental management of the world.

THE ANCESTRY OF THE EUKARYOTES

The early Proterozoic world was dominated by bacteria. Eubacteria occupied the shallow seas and muds. Archaebacteria were there too, and also in hot springs and hydrothermal systems. Among these organisms were the ancestors of the eukaryotes (see Chapter 5). Sometime, possibly by the early or mid-Proterozoic, they evolved to become the distinct third dynasty of life, the ancestors of all the higher organisms.

Whether there was only one first line of eukaryotes or several, we do not know. Today, the eukaryotes are split into several lineages: the animals, the plants, the fungi and one or more lineages of simple organisms, the protists. All these organisms share certain characteristics, but the divisions between them run very deep. They all have their DNA mainly in a nucleus – prokaryotes do not have a nucleus – but some DNA is held in separate structures known as organelles. These organelles include the chloroplasts in plant cells, which carry out photosynthesis, and the mitochondria that exist in the the cells of nearly all eukaryotes, plants and animals, and are the powerplants of those cells. The nuclear DNA is passed down from generation to generation by sexual or asexual reproduction, but the separate DNA in the organelles is usually passed to the offspring from the mother only, not the father. This separate DNA is

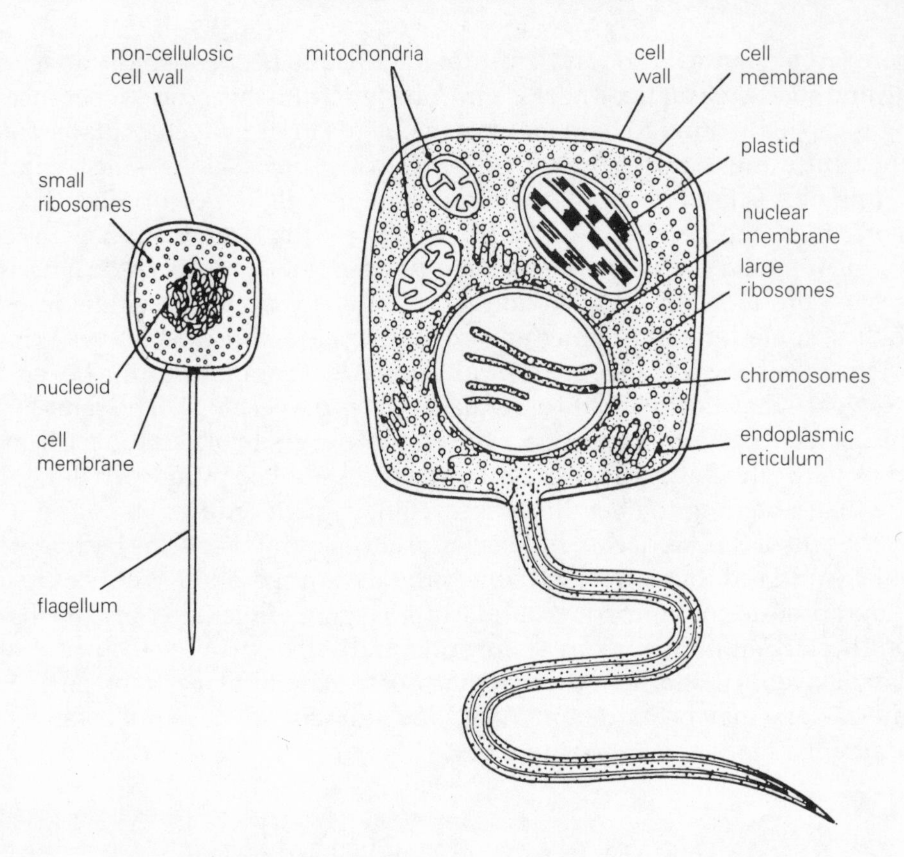

Figure 7.1 A prokaryote (left) and eukaryote cell (right) (from Margulis L. & J.E. Lovelock 1982. In *Life in the Universe*, J. Billingham (ed.). Cambridge, Mass.: MIT Press).

probably derived almost entirely from symbiotically incorporated bacterial ancestors, while the nuclear DNA seems to have originally had some archaebacterial characteristics.

With some imagination, the ancestry of the eukaryotic cells may be reconstructed, possibly to an original fermentative host organism, perhaps an archaebacterium. This organism acquired a purple non-sulphur bacterium that became the mitochondrion, and also a cyanobacterium that became the chloroplast. Margulis and Sagan suggest that the original mitochondrial ancestor may have been a predator that accidentally took to living in peace with its host, just as, on a different scale, humans have useful symbiotic bacteria in their gut in contrast to the bacterial infections that make us sick. A reverse process may have occurred during the origin of the ancestral chloroplast. Margulis has suggested that another incorporation also took place, of a bacterium similar to organisms known as spirochaetes. This symbiosis, if it occurred, could have produced the whip-like tails (undulipodia) that allow simple eukaryotes to move. Higher organisms may have constructed their nervous systems from this spirochaete basis. If this idea is correct (and many biologists do not accept it), then it is to those symbiotically associated spirochaetes that we owe our brains, and through them that we read this page.

During the aeons since these ancient symbiotic marriages, much of the essential DNA originally in the chloroplasts and mitochondria seems to have been moved from the organelles to the cell's nucleus, but enough still remains to carry out the essential functions of the organelle. It is by studying this remaining DNA that we can, for instance, deduce that the photosynthesizing eukaryote is simply a new home for the old cyanobacterium: the DNA in the chloroplast is closely similar to that of the bacteria. By this symbiosis the cyanobacteria have been able to colonize the land, create the Amazon forest and persuade the authorities at Wimbledon to grow a fine green patch of lawn for tennis. New plants are simply old cyanobacteria writ large. Similarly, the animal mitochondria that were once bacteria are now passed down from generation to generation by reproducing animals. Each human being receives slightly more from the mother than from the father, because the mitochondrial information is purely maternal. For the symbiotic bacteria now incorporated in eukaryotes, the arrangement has great advantages. Today, encapsulated within the animals and plants, they walk the Earth, and inhabit the skies and the ocean depths. We have gone to the Moon carrying with us in our guts our microbial eubacterial and archaebacterial cousins. A human being is simply a space suit for an assortment of bacteria. This, of course, is looking at matters from the bacterial point of view: the arrangement is also to the benefit of the eukaryotes, which could not learn the tricks such as photosynthesis afresh.

REPRODUCTION AND DEATH

Unlike bacteria, eukaryotes are created as male and female. Eukaryotes do not necessarily need sex, yet they generally have it. Even relatively recently evolved plants, such as tulips, dandelions and potatoes, can spread without sex, and many animals can reproduce asexually. In theory, human beings too can be cloned. Sexual reproduction would appear to be disadvantageous to a species: it means that part of the population, the male part, is unable to reproduce. In a competitive setting, this is very dangerous. Fewer offspring can be born and the sexual organism runs the risk of falling behind in the reproductive race.

In the longer run, sexual reproduction offers a different way of rearranging the genetic material and thereby evolving new forms more efficiently than in asexual reproduction. Sexual reproduction creates competition between males and also provides a mechanism for correcting genetic errors in organisms. In the short run, however, rapidly breeding asexual organisms appear to have the advantage. Possibly, species that reproduce sexually may be sacrificing short-term advantage for long-term gain. It is almost as if in climbing to the evolutionary heights each generation has to cross a dangerous ravine. Why on Earth does sexual reproduction survive? The answer is not really yet known. Yet many of the eukaryotes reproduce sexually, and many of them have roughly equal numbers of males and females. In a few species, mostly vertebrate, there is even the odd arrangement that males may match females in physical size and status.

It is worth considering an analogy here, leaving science for a moment to consider the evolutionary complexity of sex and death. Many years ago I studied the introduction of advanced farming methods in a rural area in Africa. Most of the people were poor, but one local farmer had become very successful. He had managed to acquire no fewer than six wives. Each wife laboured in the fields for him, generating income, and each wife produced children to her maximum biological capacity. The daughters so produced eventually went off in marriage, the bride money paid to him by each new husband bringing in the price of ten cows; the sons helped in the fields or sent back money from the towns. As he became richer, he married more young wives, who produced more children to make him richer, and so on. In contrast, his monogamous neighbours continued in poverty.

The man had become rich: for him life was pleasant, and he had had many sons. Eventually, a few generations hence, the economy of the area will crash as the population becomes too great for the local resources. But when it does, even if many people starve to death, by then his descendants will greatly outnumber those of his near neighbours: however many people die, his descendants maintain their advantage. After several cycles of boom and bust, the neighbours will have either been exterminated or become polygamous themselves in an effort to compete. Only the polygamous man passes on his genes, eventually. The fitter genes dominate. The more wives per man, the better to survive. The ratio of females to reproducing males rises. The apparently logical end, of course, is to eliminate the man almost entirely.

Yet the elimination of the male has not happened in the eukaryotes. In our human societies, in the West, the church has historically abhorred polygamy. In economic terms, the nuclear family allows society to improve the individual lot of the females and children, and in consequence, the collective lot. But such altruism can be imposed on humanity only with the greatest difficulty: the polygamous peasant plutocrat was not at all happy with the claims of his church that he desist from the ways of Solomon, and listen to the preaching of St Paul that women, different races, and even slaves have rights: 'There is neither Jew, nor Greek, slave nor free, male nor female, for you are all one'. Unlike humans, protozoa do not philosophize: why do simple organisms act with apparent altruism, or interest in the future, if it means denying the interest of the present generation?

A possible answer is that sexual reproduction was accidentally locked into the eukaryote way of life at some early stage, when competition was light. Then the short-term danger of sexual reproduction was small and the longer-term advantages of sexual gene-shifting were discovered. In this hypothesis, the ancestral organism, having moved briefly to a sexual way of reproducing, was never able to return to asexuality, and became locked into the system, just in the way that the top row on the keyboard of a word processor still keeps the ridiculous QWERTYIOUP arrangement of keys. Somehow, though, this argument seems fake. Many eukaryotes are able to reproduce either sexually or asexually, yet they do not abandon sex. Why?

Perhaps a part of the explanation lies in another aspect of the eukaryote conditions: death. Bacteria die only when they run out of food or when they are irradiated or subject to some catastrophe. Primitive eukaryotes, such as amoebas, do not necessarily die either. Yet many sexual organisms seem to die deliberately, as if death is programmed into their existence. Death removes the previous generation; death allows the new products of the sexual genetic shuffling to dominate. Seen from the competitive stance of the gene, death is a marvellous survival tool; although each living individual may not be so keen on it, death is much to the advantage of the genes that pass on to the children, which do not need to fight the parents for scarce food.

Interestingly, there is a correlation between the degree of predation suffered by a species and its lifespan and reproduction rate. Animals such as mice and rats, and many insects, reproduce rapidly and are as rapidly eaten by predators. Each individual has many offspring: a few survive to reproduce. Lifespans are short. In contrast, where there is little competition, species have few offspring and, instead, the individuals have long lifespans. The energy of the species is invested in a few long-lived individuals, not in a host of brief candles.

The birds of New Zealand are a good example. With no natural predators, the world's largest parrot, the kakapo, probably lives as long as a human being, and has few chicks. Today, on the New Zealand mainland, only a few males are known to survive in the wild: the unprotected birds couldn't handle cats. Part of the reason for their tragedy is that they are flightless and ground nesting, but so are other more successful birds: kakapos breed too slowly to make up for the loss.

In prehuman New Zealand, the long life and slow breeding of the kakapo may have evolved because advantage was gained by those genes that invested greatly in an individual body that was likely to survive for many years, with few competing offspring being born. A parallel logic holds in human society – in comparison to the poorer nations, Western nations, with few children, and long individual life spans, are better able to train their population and can use both male **and** female in the workforce. In consequence, the rich nations abandon the short-term advantage of a huge population, but are rewarded by being less vulnerable to catastrophes such as famine or disease, and more able to train their population to create and face the future. In maximizing the living biomass of a species, there is a trade-off between lifespan, death and offspring.

Many eukaryotes seem to have had death programmed into their make-up, perhaps because there is a selective advantage to the progeny of those parents that die after useful lifetimes but before they compete with their children. Whatever the cause, sexual reproduction must be an asset, because it is maintained.

Bacterial genetic exchange is very different from this. As mentioned earlier, although we can identify individual strains of bacteria with distinct characteristics and which reproduce consistently, bacteria are also able to incorporate genetic information from other bacteria. Occasionally, recombination can occur between apparently very distant bacteria, one type learning from another. It is as if a herd of hippopotami learned to fly by developing wings after a passing romantic acquaintance with an ostrich, which had itself regained the ability to fly after bumping into a swallow. Specific abilities can be acquired by one strain of bacteria through the selective transfer of genetic information from other strains. Most probably, the basic tools of bacterial life, for instance being able to fix nitrogen, evolved once, and since then have been selectively acquired by strains of bacterium that could use the ability.

In contrast, our sexual method of genetic transfer is unable to do this: a hippo calf cannot acquire feathers because its mother met an ostrich. Eukaryote groups evolve to become distinct entities, separated from and infertile with even close genetic cousins. They cannot, usually, learn evolutionary tricks from other organisms. Of course, there are exceptions to every rule, and genetic material can, rarely, cross species barriers, even in higher eukaryotes.

THE EARLY EUKARYOTES

We do not know when the eukaryotes began, but by the late Proterozoic the oceans must have been populated by countless numbers of protists, as are today's seas. Protists are simple eukaryote organisms which form part of that kingdom of life that includes simple single-celled eukaryotes such as amoebas and algae. We have no direct fossil record of when protists first evolved. Some microfossils are preserved in rocks known as the Gunflint chert of north-western Ontario, which are about two billion years old, but these microfossils appear to be prokaryotic, not eukaryotic. Simple unicellular protists are most unlikely to have been preserved in an identifiable form, and so we cannot even say that they are younger than the Gunflint chert. Possibly, molecular 'clocks' can be used to estimate the time needed to carry out the genetic changes in the early development of the protists. There are also an array of fossil markers: small obscure fossil structures. However, in our present state of knowledge all we can say is that the critical symbiotic amalgamations that produced the

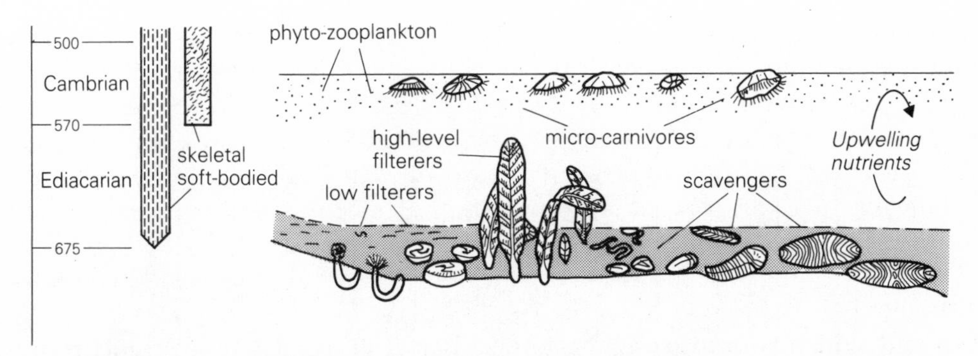

Figure 7.2 The Ediacarian soft-bodied metazoan community (courtesy of Paul Copper).

eukaryotes most probably took place in the Proterozoic, possibly around 1.5 billion years ago, and almost certainly before 850 million years ago. At least two and perhaps more symbiotic associations occurred.

The early protists were probably unicellular organisms similar to many modern plankton, though in some cases bigger than today. One guess is that the ancestral line was photosynthesizing, possessing incorporated mitochondria, chloroplasts, and perhaps spirochaete-derived tail-like structures. Some protists have the ability both to use photosynthesis and to ingest food: thus they are neither plants nor animals but have properties of both.

From our eukaryotic point of view, life can be divided into five kingdoms. At the root is the bacterial, prokaryote kingdom. Out of the prokaryotes stem the four eukaryote kingdoms: first the protists, or protoctista, then the three familiar kingdoms of plants, animals and fungi. Not counted in these kingdoms are the viruses and fragments of genetic information that cross from organism to organism. This division is from our point of view. Were bacteria able to classify themselves, they would recognise the three dynasties of eubacteria, archaebacteria and eukaryotes, with various smaller subdivisions such as plants and animals within the eukaryotes. But it is for ourselves that we classify, and the division into five kingdoms serves our interest well.

We can picture the seas, say, 800 million years ago, filled with an assortment of bacteria and eukaryotic plankton and algae. Life covered the globe – probably even the land surface had been colonized by bacteria. The ecosystem was simple compared to today's diversity of organisms, yet the bacterial community and simple eukaryotes used the same basic tools – chemical reactions – upon which the modern biosphere depends. Perhaps some of the eukaryotes were living in colonies of unicellular organisms; possibly some, like amoebas, had abandoned photosynthesis to become 'animals', living off the bodies of their relatives; others may have remained as plants, like algae, unable to ingest material in order to concentrate on photosynthesis. From these asexual beginnings came the complex, often sexually reproductive eukaryotic ecosystem that inhabits the Earth.

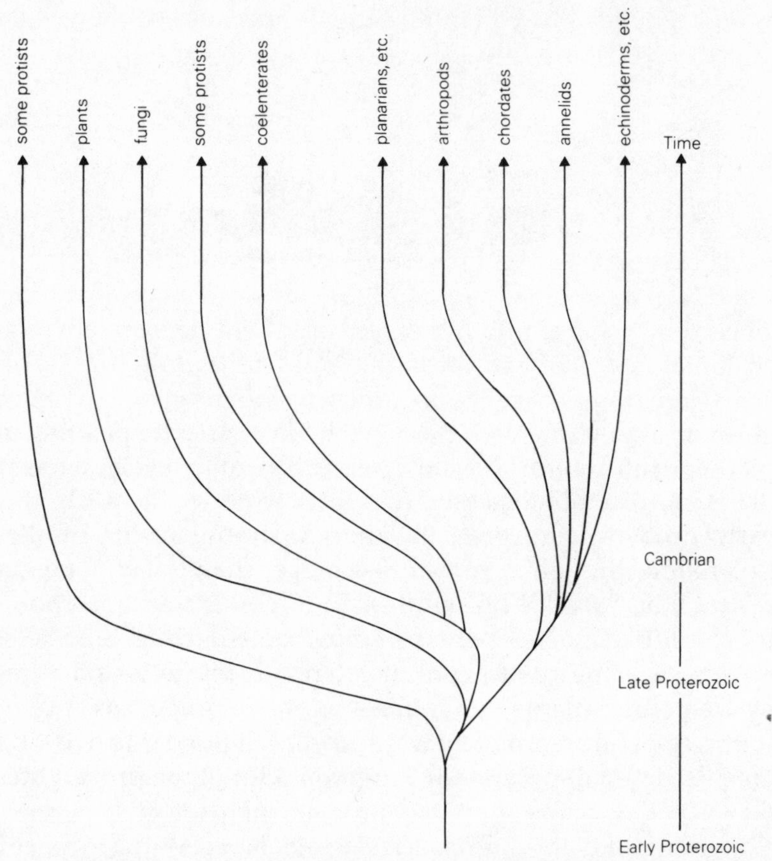

Figure 7.3 The relationships between the eukaryotes. These lines of descent are to a large degree based on evidence from ribosomal RNA. Protists include single-celled organisms; the term Protoctista is also used, to include single-celled organisms and related multi-celled forms. Of the more complex organisms, examples include yeast (fungus); sea anemone (coelenterate); insects (arthropod); humans (chordate); earthworms (annelid), and starfish (echinoid). Evolutionary tree is derived from RNA sequences, especially work by Field and others, *Science* **239**, 748–53.

Once a property or an ability is given up in a sexually-reproducing organism, that change cannot be reversed. Any bacterial line, having abandoned photosynthesis, could in some later generation re-acquire the trick and the tool-kit by absorbing genetic material from a bacterium that still knew how to do it. But sexual organisms do not usually have the ability to collect small packages of DNA information from a neighbour of a different species (although there are minor exceptions to this). Once a tool-kit has been abandoned, it is lost forever; if it is needed again, it must be accidently reinvented, or the line of organisms runs the risk of extinction. When a line of higher eukaryotes specializes, it then has to evolve in a particular direction. It is set, forever, as a plant, or animal, or fungus. In a sense, human society, which trades information, is closer to the bacterial model. We can learn, forget and relearn. We can degenerate as societies and rebuild. But when the protists specialized, they began the path that led to the modern world. There could be no turning back, although on several occasions there have been catastrophic halts.

THE METAZOA

Single-celled eukaryotic organisms can, on occasion, form colonies of many cells. Sometimes reproduction may not involve physical separation, so that a clump of cells may form; in other cases cells seem deliberately to associate to form a colonial body. But towards the end of the Proterozoic, nature went beyond this and groups of cells appeared, organisms with a definite plan, each cell performing a specialized function in the interests of the whole (and of itself). Some cells even die in the interests of the whole group; just as we are constantly and deliberately sloughing off unwanted skin cells. We do so that the body may live and reproduce (and produce more skin cells, if the matter is seen from the skin's point of view).

Consider a piece of wood, a foot long. Imagine a rule that states that at the ends of any piece of wood a cross-piece must be nailed in at right angles, and that the cross-piece must be exactly half the length of the first piece. Two six inch cross-pieces must be nailed on, one at each end of the piece of wood. Now there are four ends: four three-inch cross-pieces must be nailed on. And so on, *ad infinitum*. Eventually, a very complex structure is produced. All we started with was some wood and a simple law: there was no grand plan or blueprint. Yet the wood and the law produced a complicated and rather attractive pattern, like a snowflake.

Organisms are much more sophisticated. Their law of growth appears to change as the organism grows, so that when a certain stage is reached the organism differentiates to make something different. More than that, the apparent change in the law seems to be controlled by the local setting: for instance, in the H-shaped example above, the original law could specify that if ever a new piece has to be put in closer than one inch to an earlier piece, the new piece must be put in not at right angles but at 60°. The pattern resulting would be substantially different. Somehow, as growth takes place in an embryo of an organism such as a plant or an animal, the cells seem to be signalled when to specialize as leaves, or bark, or hands or toes. The first cells of an embryo can become anything, later cells take up distinct roles; a cell that has differentiated to become part of an arm cannot perform the functions of a liver.

In traditional classification systems, single-celled eukaryotic animal organisms are called **protozoa**, while multi-celled animals are termed **metazoa**. The slime mould amoeba is a protozoan organism. Normally it lives as independent cells, feeding on bacteria and yeast, and reproducing by division. If the food supply is short, the amoebas stop reproducing and collect together.

They form an extraordinary object about 1–2 mm, (a sixteenth of an inch) long, a wormlike slug that slithers about. Eventually, the cells differentiate and the slug turns into a body that releases spores which can germinate later when conditions are better.

Several simple organisms can form fruiting bodies. In contrast, sponges are multicellular animals which have several specialized cell types. However, when a sponge is put through a mincing machine and broken up into cells, the individual cells can collect to form new sponges. Sponges are at the start of a trend towards complexity in organisms, in which cells depend on each other in a body. More sophisticated multicellular animals, such as people, cannot reassemble themselves when minced up. They have highly ordered and complexly interdependent hierarchical structures. These are the **metazoa**. They first appeared on Earth 670 million years or more ago, when our first record of complex animals begins.

THE END OF THE PROTEROZOIC: THE EDIACARIAN RECORD

About 600 km north of Adelaide, in South Australia, is a range of low hills in hot dry country. In these hills, the Ediacara Range, are outcrops of a geological unit known as the Pound Quartzite, which was originally laid down as sand in shallow intertidal waters. The sand was deposited about 650–700 million years ago and is overlain by considerably younger Cambrian sediments. The sand has now turned to rock, a quartzite, and imprinted in the beds of the rock are the shapes which many geologists interpret to be those of some of the animals that lived in the Ediacarian Sea.

These shapes give us what is perhaps our best picture of late Precambrian life. Imprinted on the rocks are various almost-familiar forms. These can be interpreted as jellyfish, worms and so on. From these imprints, geologists have reconstructed a picture of the marine community at the end of the Proterozoic. Jellyfish swam in this sea, large leaf-like animals rooted themselves in the sea bottom, while segmented worm-like animals swam through the water or crawled across the surface. None had hard parts though, so most would have been broken up by water currents as they rotted after death, leaving little behind. But fortunately, the imprints of some were preserved – today such an animal, if it died, would immediately be disposed of by carrion-eaters. Indeed, one palaeontologist has pointed out that the late Proterozoic was the only time when, apparently, chunks of meat rained down onto the sea floor and did *not* get eaten. There was a free lunch, but nobody to eat it. Relicts of similar organisms have been found in Charnwood Forest in England, in Namibia and elsewhere. By the end of the Proterozoic a diverse animal community had developed, a whole range of specialist organisms forming a new ecological web.

We see here the beginnings of a new order. The new specialized multicellular organisms, the metazoa, evolved to occupy ecological niches; as they did so, they increased the diversity of the biosphere, and in their diversity they also created new niches for yet other new animals to fill. Evolution is a spiral, interacting process. An organism arrives to fill a vacant spot and, simply because it exists, it helps to create a new vacancy.

Consider how towns grew in the American West. Originally, a few people clustered together, perhaps to buy or ship agricultural products and to distribute and sell things (such as drink) to local cowboys and farmers. Then a physician would come (perhaps because of the drink). People would have babies and perhaps a schoolteacher would be needed to cater to the new population. Then a bank would be built, to keep the money of the society. Then, eventually, as the pinnacles of civilized development were reached, a fast-food restaurant would be needed, a take-out pizza joint and a car-wash. Most people in a town serve each other: the growth pattern spirals 'upwards'. There is constant competition between the components of the society – restaurants and shops are always going bust, almost as fast as new ones are opened – but there is also a deep interdependence of the whole fabric.

Evolution is exactly the same sort of process. In the Ediacarian the complex metazoan fabric of the modern biosphere was just beginning, adding itself to what already was an ancient and diverse bacterial substratum of life. Eaters arrived: eventually animals did evolve that were capable of disposing of those chunks of meat raining down, and the community became more diverse and richer.

THE ENVIRONMENT AT THE END OF THE PROTEROZOIC

In his book *Gaia*, J.E. Lovelock digresses briefly to recount the illustrative parable of Dr Intensli Eeger, who invented and released into nature a new organism capable of fixing nitrogen and phosphate very efficiently. In the story, the organism runs wild and removes all phosphorus from the environment: the result is that the delicate ecological balance of the planet collapses. Eventually, millions of years later, the atmosphere returns to its natural state, dominated by carbon dioxide, above a salty sea and a sterile planet. Sir Fred Hoyle invented a rather similar nightmare in his second Andromeda tale. All this is science fiction, and we hope it is unlikely (Lovelock gives Dr Eeger a second chance in a later book), as life has, we hope, probably already developed virtually every conceivable poison and damaging organism. Nevertheless, the story serves a purpose. It illustrates the central tenet of the Gaia hypothesis: that the environment is biologically controlled. The bacteria may have begun this control, but the evolution of plants and animals made it much deeper, more pervasive and more sophisticated.

Carbon dioxide is obviously biologically managed by plants and animals. Recall that on Venus, carbon dioxide forms most of the atmosphere, while on Earth it is an essential but trace component. About as much carbon dioxide is stored in Earth's limestones and dolomites as is free in the air of Venus, and carbon is also stored in the Earth's mantle as carbonates, methane and even in diamonds. The carbon dioxide level in the air today is closely controlled by the activities of eukaryotes, in the balance between plant uptake and the emission of the gas by animals and plants. This eukaryotic control began in the Proterozoic, starting with the evolution of plankton.

The arrival of the eukaryotes would have had important effects on the oxygen content of the air. As their influence as producers and consumers became felt, the rules governing the oxygen balance of the surface may have been rewritten to some extent, even though the eukaryotes, through symbiosis, were carrying on the basic metabolic patterns set up by the bacteria. On the modern Earth, oxygen forms more than a fifth of the air. Most scientists (but not all – see the discussion on the Archaean atmosphere in Chapter 5) believe that, over the aeons, life has slowly and steadily acted to build up the oxygen levels in the air, to reach this modern level. In this model, originally, very little free oxygen may have existed. By the early Proterozoic the actions of cyanobacteria had placed enough oxygen in the air to produce red, or oxidized, sediments where material was exposed to the air. Then the protists incorporated cyanobacteria and learned how to produce oxygen; animals developed to consume the oxygen. The alternative model (see Chapter 1) is that the air has been oxygen-rich ever since the Earth lost much hydrogen to space early on by the action of ultraviolet light on water vapour at the top of the atmosphere. Loss of hydrogen from water leaves free oxygen in the air. This process would have been very important in the early Earth (see Chapter 1), but is not significant today because of the atmospheric cold-trap. In either case, the production–consumption cycle is probably extremely old, as Archaean bacteria were essentially doing before what plants and animals do now, although when the eukaryotes arrived the rate of cycling would have increased. While both organic and inorganic production of oxygen occur, today the management of oxygen is biological. But if so, how is the level of oxygen, and its ratio to nitrogen in the air set?

We do not know the answer. Today, nitrogen acts to dilute the oxygen and to inhibit fires. Humanity has learnt, at a price, the importance of nitrogen in preventing fires. The early Apollo spacecraft were intended to operate with a cabin atmosphere of oxygen, without the other gases in the air. This made good engineering sense since it allowed a low interior pressure in the cabin, but a disastrous cabin fire cost three astronauts their lives. Oxygen by itself is immensely dangerous: it burns. Add nitrogen, and conditions are much safer.

In the Proterozoic there were no land plants, so the maintenance of oxygen levels by fire would not have been possible. Presumably there was some trade-off between, on the one hand, the biological processes of producing and consuming organisms, and on the other hand, inorganic processes, such as the oxidation of iron in ironstones or the reduction of seawater as it passed through mid-ocean ridge hydrothermal systems. While carbon was being fixed in limestones, dolomites and in the sea floor, huge amounts of oxygen were probably removed from the air and returned into the mantle via the subduction of oxidized sea floor. Oxygen would also have been removed by deposition in iron-rich sediment.

In comparison with the enormous amounts of carbon and oxygen sequestered in this way over the millions of years, and also in contrast to the sum total of matter cycled biologically between the atmosphere and living organisms, the 'standing crop' of gases – their steady levels – in the atmosphere is small. We can only guess the composition of the Precambrian air. The geological evidence is ambiguous. Most of the preserved rocks that could tell us, such as fossil soils, have been altered by weathering, or metamorphosed, or changed by later fluids moving through them, so that the information has been degraded. On Earth today there are abundant oxygen-poor settings, such as muds on the bottoms of rivers and pools: there are also oxygen-rich environments, where we live. Both can be preserved in the sedimentary record, and the Proterozoic record seems to indicate much the same, with abundant oxygen-poor deposits, but also some that appear to have been laid down in oxygen-rich settings.

Perhaps we shall be able to reconstruct the oxygen levels of the Proterozoic air only by experimenting with the complex interactions and needs of simple microbial organisms: they may have set the levels to suit themselves. Possibly the oxygen level was not influenced by them: they may simply have allowed the mid-ocean ridges and their hydrothermal systems to do the job. But life is an interfering busybody: perhaps the ironstones were precipitated by organisms that made a living by removing some of the oxygen released by the actions of the cyanobacteria and the chloroplasts.

In our mind's eye we can picture the late Proterozoic planet, bacteria fitting and filling every possible niche available to them, including stromatolites making oxygen in shallow water, the oceans teeming with plankton, and the more complex newcomers – jellyfish and worms – dominating the visible ecology. On land there were perhaps a few lichens and fungi, possibly even worms in damp places. Above, the air was probably mainly nitrogen, with some oxygen, carbon dioxide and water vapour, biologically managed, though we do not know at what levels. Earth was a blue–brown planet, with wide seas, bare land and some white clouds, all being shaped by life.

FURTHER READING

Glaessner, M. 1984. *The dawn of animal life*. Cambridge: Cambridge University Press.
Holland, H.D. 1984. *The chemical evolution of the atmosphere and oceans*. Princeton, NJ: Princeton University Press.
Margulis, L. & D. Sagan 1986. *Microcosmos*. New York: Simon & Schuster.

PART 3

The spread of the eukaryotes

8

The Early Palaeozoic explosion and its aftermath

THE SURVIVAL OF THE FITTEST

In the middle of the eighteenth century, war was the business of gentlemen and their hirelings. It was fought by relatively small armies which were manoeuvered across Europe and the world as chessmen across a chessboard. War was a polite, cruel business: the Huns and the Vandals had been forgotten. The victories of 1759, which set out the map of the modern world, were won by small forces in tiny decisive battles. Later, after the French Revolution came the democratic invention of citizen armies – masses of men trudged across the continents to do the duty of battle. Out of this came the machine gun and barbed wire, the slaughter of the American Civil War and the trenches of Flanders. For each new weapon of offence was an equal weapon of defence, and millions faced each other in a stalemate of ignorant armies clashing by night or, as in 1940, a desperate struggle for survival. We have now reached a splendidly final state where our best defence is the assurance that we can destroy the enemy even as we ourselves are destroyed.

This spiral of battle, the development of better and better weapons of offence and defence, is not a human invention. It began aeons ago. It may even be that our first eukaryote ancestor was a product of a battle between one bacterium and another – though that battle ended not in victory but in compromise with the combatants developing a way of living together. Most wars, whether human or animal, do not end this way. In recent history, few wars have ended in a compromise and symbiosis. Usually, the loser is killed. Or, if defeat is not absolute, the loser retires to some dark corner of Carthage to fashion new weapons, until final victory or deletion. Animal wars frequently take this pattern – the fittest survive and the loser becomes extinct. But bacterial and viral wars can be more subtle, with the exchange of information between the competing organisms.

Natural competition can be more subtle than simple battle. I once watched a cheetah and her cubs amusing themselves in the late afternoon by playful battle amongst each other: a few metres away stood an unconcerned wart-hog and her clutch of piglets. Supper? No. Presumably the cheetah family had eaten, and somehow the hogs knew. More generally, the predator depends on the survival of the prey, because if the prey becomes extinct so does the predator. The piglets had no immediate cause for fear.

Human beings, too, do not always fight. Sometimes to fight is seen as not being worthwhile, not gaining an advantage. Pigs bring to mind the tale of the last great war between two major democracies, the celebrated 'Pig War' between the United States and Canada in 1859. A Canadian pig raided an American garden. The United States 9th Infantry was sent to the rescue. The great military machines of the British Empire and the United States of America braced for conflict. The troops met on the battlefield. The military demand blood in any good battle: the victims, so the myth goes, were the pigs, which provided a series of excellent barbecues for the soldiers of both sides. This story is now entangled in legend, but in general the folk-tale illustrates truth: democracies care too much about food to fight. And so, often, does nature. It can be more productive if a predator does not always kill. Nature is not always bloodied in tooth and claw. Lions spend much of their time mating, not killing; cheetahs 'allow' their prey to breed. The effect is to make dinner sustainable . Predator and prey co-evolve.

War probably started in the Archaean, and in earnest in the Proterozoic. There is little record of these anonymous conflicts. Most fossils are not the weapons of attack but the tools of defence. The most permanent record of 1940 in England is not the fighter aircraft, very few of which survive, but the lines of concrete pillboxes strung across the countryside as preparations against invasion: at Pevensey Castle on the south coast of England the defensive fortifications record almost every conflict from Roman times into the period of the atomic bomb. Similarly, the history of life is best recorded by the defensive works, the fortifications created by organisms. Some attack weapons, such as teeth, do survive, but most fossils are of defensive structures. In the history of life the most striking change in the quality of the documentary record comes with the evolution of fortification: the development of hard parts. This advance marks the beginning of the abundant fossil record, because hard parts stand a good chance of preservation when buried in sediment.

Suddenly, at the beginning of the Cambrian, animals began to defend themselves. The animals being preyed upon surrounded themselves with walls of concrete or with a hard organic material, like the casing of lobsters. The ferocious world of eat-or-be-eaten had begun. Why hard parts arrived so suddenly we do not know – it is one of the mysteries of palaeontology. Perhaps it was something as simple as the acquisition by a single predator of a good set of eating equipment, so provoking defensive responses in other species:

perhaps something else happened. Whatever the cause, the seas became violent and vicious environments, and have remained so ever since. Death from old age is rare: most organisms get eaten.

Hard parts make up the bulk of the fossil record. But skeletons alone do not allow us to imagine the whole animal. The story of life in the Cambrian, the early history of the animals, is wonderfully written in an extraordinary fossil locality that records soft parts too.

THE BURGESS SHALE: LIFE IN THE CAMBRIAN SEA

The Burgess shale occurs in the Canadian Rockies, not far from the little village of Field on the Kicking Horse Pass. Here, on the slopes between Mount Field and Wapta Mountain is a geological site so important that it has been designated a World Heritage Site by the United Nations. During the middle Cambrian a huge reef built by algae grew above the strata preserved here. Nearby, in the mud, the bodies of dead organisms were preserved. What is extraordinary about the site is that fossils of soft parts were preserved, not just the hard parts. As a result, we have an almost unique snapshot of life in the Cambrian Sea, predators and prey together, soft organisms and the harder-shelled creatures and crustacea.

The Burgess shale records a whole community of life. Among the plants, there are delicate remains of seaweeds and algae, in masses of broken fragments. Possibly some of the algae were swept in from the underwater cliff edge that stood above the muds, which occasionally slumped off and carried algae downwards. No land plants existed in the Cambrian as far as we know, so the work of photosynthesis that maintained the planetary atmosphere was done by algae and the oceanic plankton.

The animal community preserved in the Burgess shale is diverse and abundant. The names of these animals are not as well known as the names of the dinosaurs, but they are scientifically more important than *Tyrannosaurus rex*, for the progeny of a few of them still live today. Each organism is equipped for warfare in some way, either for attack, or defence, or hiding. The fossils of the Burgess shale include representatives of many of the major divisions of animal life that we know today. Associated with them, though, are other groups that later became extinct – the failed branches of the tree of life.

The commonest organisms in the shale are the arthropods – the jointed animals that today include insects, spiders, crustaceans and the like. These animals have an outside skeleton of tough organic material – like beetles and grasshoppers – or they can reinforce their body with calcium carbonate, as in parts of crabs and lobsters. A common type of Cambrian arthropod is a group, now extinct, called the trilobites. Many of these are preserved in the Burgess shale, including whole bodies. Other arthropods also occur, including types which were essentially soft-bodied. One of them, *Canadaspis*, is probably similar to the animals from which modern shrimps, crabs and lobsters are descended. Another, *Sidneyia*, is broadly similar to the modern horseshoe crabs. Some of these animals, such as some trilobites and *Sidneyia*, were probably active carnivores. Some specimens of *Sidneyia* actually contain fragments of their last dinner, including pieces of trilobites – *Sidneyia* was obviously able to eat other animals with hard covers. Other animals, such as *Canadaspis*, probably ate up scraps of organic matter while others ate tiny bottom-living animals. There was thus a complex food chain in operation.

There are modern parallels to some of the Burgess animals. In 1873, the crew of HMS *Challenger* were in the early stages of their global voyage of discovery. They had landed briefly in South Africa at what was then the friendly, multiracial (blacks had land and the vote, then) and liberal self-governing Province of the Cape of Good Hope. The hospitality was excellent, the people of all races delightful and, at the little military settlement of Wynberg, so was the restorative power of a mixture of milk, eggs and brandy. Perhaps it was under the influence of this concoction – we do not know – that Henry Moseley took to inspecting the living things under an old cartwheel.

What he found under this wheel in Wynberg was an extraordinary little beast a few centimetres long. It had two antennae, a head with two eyes, jaws and a mouth, and seventeen pairs of short feet. It breathed air and, surprisingly, bore live young within it. The beast is *Peripatus capensis*, and it and its relatives are found in places as far apart as Africa, New Zealand and South America (though it is most unlikely that it could swim an ocean).

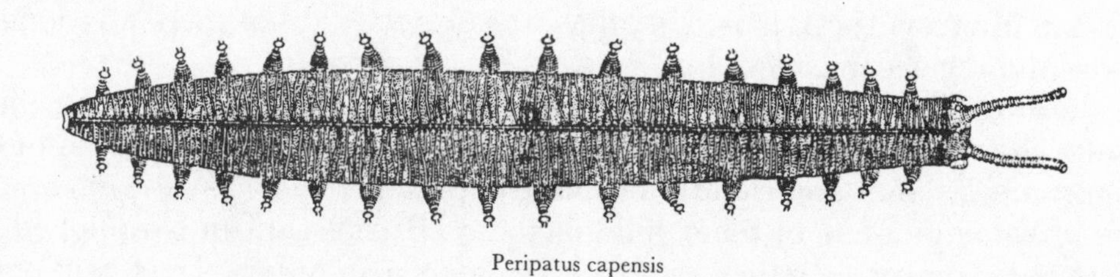

Figure 8.1 *Peripatus capensis*, 3–5 cm (1–2 inches) long (from HMS *Challenger* Report).

Peripatus has a relative, preserved for us in the Burgess Shale. The relative, called *Aysheaia*, looks almost the same, though with fewer legs. It had a thick flexible body and short legs ending in tiny claws. It was marine, not an air-breather, but otherwise it seems close to the modern *Peripatus*. Just possibly this animal was close to the ancestral stock from which all modern jointed-limbed organisms came – centipedes, millipedes and, more abundant, the insects. *Peripatus* is of course a highly evolved modern organism, but its distant

Figure 8.2 *Aysheaia*, showing the thick body and the legs. Claws are present at the tips of the legs. About 4 cm (1½ inch) long (photo courtesy of H.B. Whittington, US National Museum specimen).

ancestor *Aysheaia* may have had a very similar life style. *Aysheaia* was a primitive predator about 2.5–5 cm (one to two inches) long: a very simple animal, with a soft skin itself, and a diet of sponges (not exactly a fast-moving prey).

Soft-bodied beasts, too, are common in the shale. One, *Canadia*, seems to have been a swimmer; others lived in burrows in the sediment. Worms in this group, the polychaetes, still exist today and are common on shorelines. Another group of worms also occurs – one of this group, *Ottoia*, was armed with spines and teeth. From the contents of its gut, it seems that it ate other shells and even its own relatives: it was probably a cannibal.

Various shells exist too. Some, attached to a sponge in one sample, were brachiopods, a type of animal that still occurs today and which has played a major role in the history of shallow marine life over the past 500 million years. Brachiopods are simple animals with two shell valves that live in shallow seas and sometimes attach themselves to the sea floor by fleshy stalks. Molluscs also occur in the Burgess deposit. From them are descended our modern snails and clams. Sea lilies and sponges are present, and so is an anemone-like creature, although no jellyfish have been recognized.

What of our own ancestors? We class ourselves as chordates, animals with backbones. One of the most interesting of the Burgess animals is the primitive

Figure 8.3 *Pikaia*, a primitive chordate. The bar on the upper side of the animal is probably the notochord, a stiffened rod along the back of the animal; roughly 3 cm (1 inch) long (from Conway Morris, S. & H.B. Whittington 1985. *Fossils of the Burgess shale*. Geological Survey of Canada, Misc. Report 43).

chordate *Pikaia*. This was an organism rather like a worm, but with a stiffening rod running along its back and with segments arranged in a zig-zag pattern. It may have been related to the ancestor of the first fish, which probably appeared rather later in the Cambrian. From some of those fish were descended the amphibians, the reptiles and human beings. Quite possibly this little worm-like creature may have been very similar to our own ancestors.

The Burgess shale contains not only ancestors and relatives of modern animals: there are also some surprises. Several of the animals seem to have no obvious modern parallel. An analogy is useful here. In Henry Ford's youth there were all sorts of extraordinary motor vehicles on the road, ranging from those with engines built on modern lines, after the pattern of Daimler and Benz, through diesel engines, to steam-powered vehicles, electric vehicles and more or less any other variant that has ever been thought of. There have been cars driven by coal and even by methane gas from manure. Today, only two of those, the petrol/gasoline engine and the diesel engine, dominate the world. For all practical purposes steamers are extinct. This does not necessarily mean that a steamer is less suitable for carrying people around: the sheer weight of investment in internal combustion means that it is almost impossible to produce or sell anything else. Once the decision was made by Henry Ford and

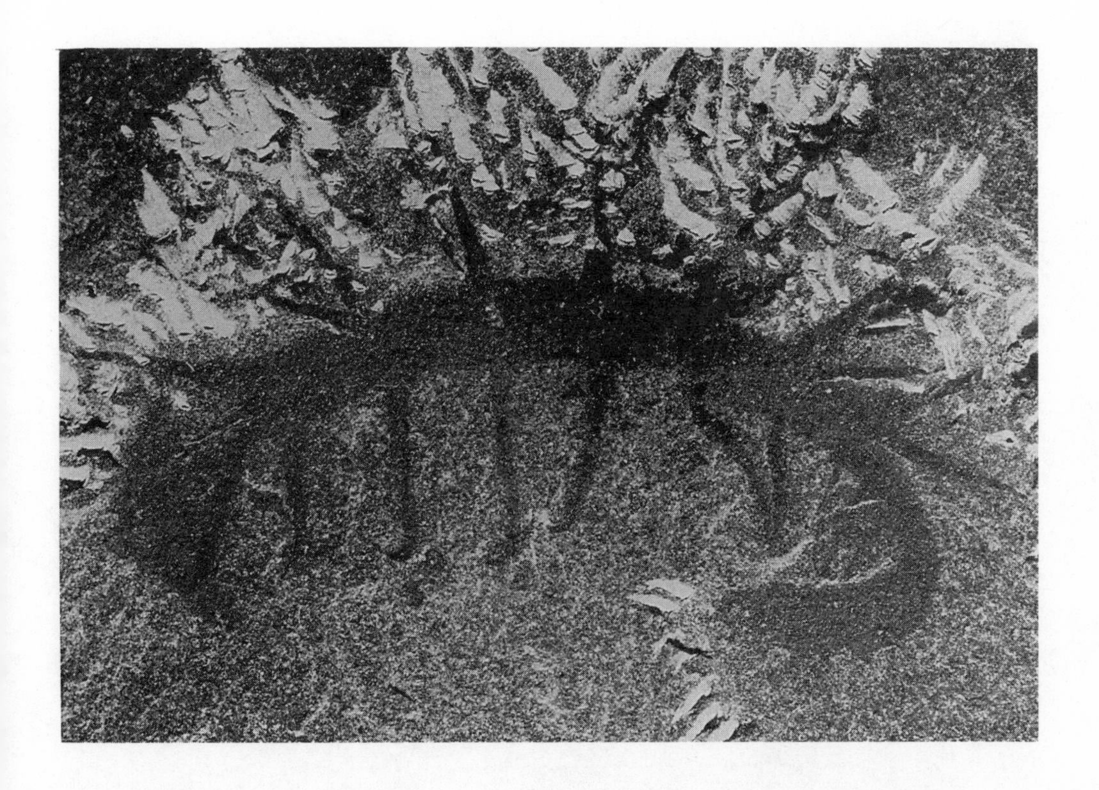

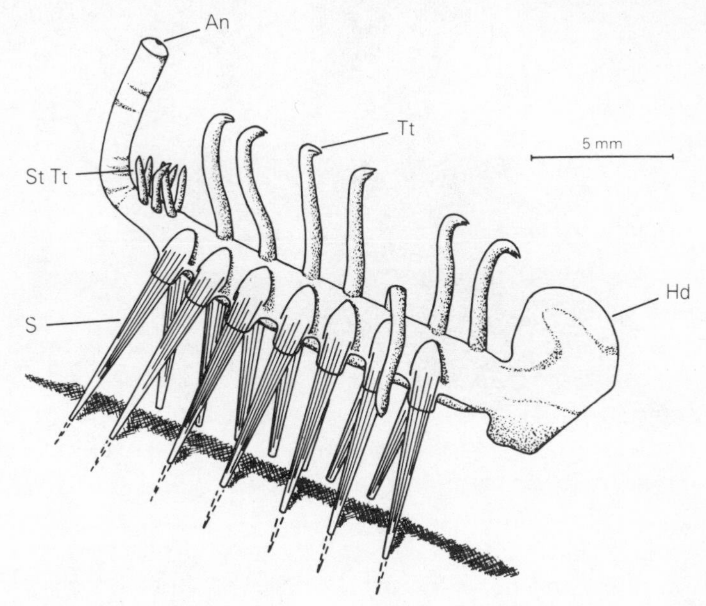

Figure 8.4 *Hallucigenia*. Top – in the rock; bottom – as reconstructed. This bizarre creature supported its trunk on seven pairs of stilt-like spines. An, anus; S, spine; St Tt, short tentacle; Hd, head; Tt, tentacle (from: (top), Conway Morris, S. & H.B. Whittington 1985. *Fossils of the Burgess shale*. Geological Survey of Canada, Misc. Report 43; (bottom), Conway Morris, S. 1977. *Palaeontology* **20**, 623–40).

Figure 8.5 *Marrella*, one of the most common fossils of the Burgess shale. An arthropod, 1–2 cm (½–1 inch) long, it sensed the environment with long antennae (a), and appendages (b) swept food towards the mouth at the back of the head. The dark stain was caused by body contents seeping out into the mud after burial (from Conway Morris, S. & H.B. Whittington 1985. *Fossils of the Burgess shale*. Geological Survey of Canada, Misc. Report 43).

others to adopt the Daimler-Benz pattern because it was better at the time, the other kinds of automobile died out.

The Burgess shale has several examples of animals which do not have modern descendants, though they must have competed well at the time. One was *Wiwaxia*, which was probably one of a fairly successful group, but has no close modern relatives at all. It was covered with scales and probably ate organic debris off the sea floor. Another was *Hallucigenia*, a bizarre creature that carried itself on stiff spines. The most impressive of all the Burgess animals was *Anomalocaris*. This beast was the battleship of the Cambrian sea. It had ferocious jaws, large compound eyes and may have been up to 60 cm (two feet) long – the largest known Cambrian animal. Yet it has gone, leaving no known descendants.

We can see the beginnings of our modern world in this collection of fossils. Here was a complex food chain, from algae and seaweed, through assorted herbivorous animals, to carnivores. Among them were the ancestors of virtually all modern organisms, including ourselves, and also a variety of animals that apparently were successful then, but with no descendants today. Collectively, the animals formed a complex network of food users, each part of the ecosystem competing with the others, yet also, in effect, co-operating to manage the food chain.

The Burgess shales contain fossils of many groups of life, such as the arthropods, the brachiopods, the molluscs, and the ancestors of the chordates – our

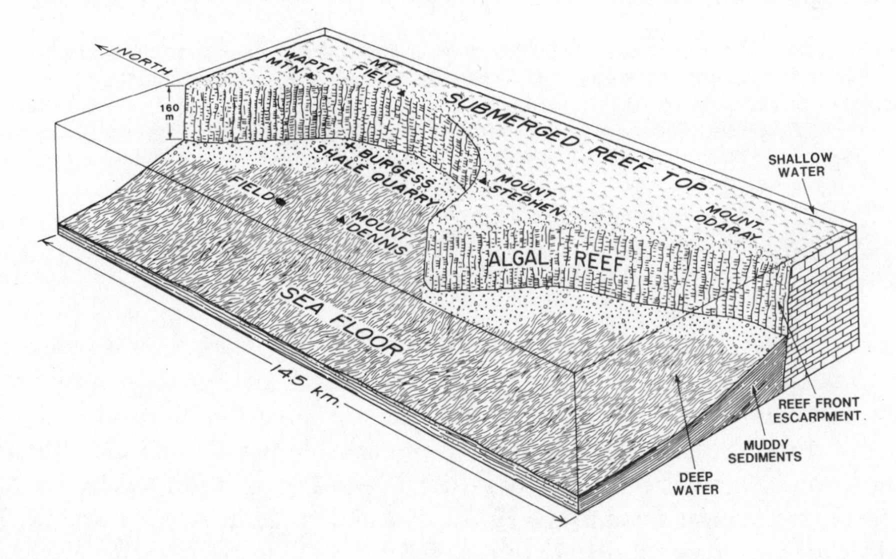

Figure 8.6 A reconstruction of the deeply embayed Middle Cambrian reef escarpment, below which the Burgess shale accumulated as muds. Near Field, on the border between British Columbia and Alberta, Canada (from Conway Morris, S. & H.B. Whittington 1985. *Fossils of the Burgess shale*. Geological Survey of Canada, Misc. Report 43).

line – that ever since have filled the planet with animals. But there were other groups, that, when the shales were laid down, were seemingly also important. These groups became extinct. Small changes, perhaps, might have made history go the other way – it could have been that our ancestors became extinct, and the 'oddities' of the Burgess shale spread down through history.

Figure 8.7 Reconstruction of some of the Burgess shale animals living on, above and in the muddy sediments being deposited at the foot of a submarine cliff (see Fig. 8.6). Animals include sponges (20, 22, 25); brachiopods (7); molluscs (16) and similar animals (4); worms (1, 2, 3), arthropods (6, 10, 11, 13, 17, 18 and *Marrella*, 15, *Canadaspis*, 12, *Burgessia*, 19, and *Aysheaia*, 5), echinoderms (21) and the chordate *Pikaia* (14). Other animals (8, 9) include *Wiwaxia* (23) and the giant *Anomalocaris* (24) (from Conway Morris, S. & H.B. Whittington 1985. *Fossils of the Burgess shale*. Geological Survey of Canada, Misc. Report 43).

MORE DIVERSIFICATION

Much of the first chapter of Darwin's study on *The origin of species* is concerned with domesticated pigeons. Wild pigeons still exist, of course, as any statue knows, but man has chosen to breed selected stock, in order to produce a wide variety of domesticated birds capable of remarkable feats. On 5 October 1850 one of these birds was released by a Royal Naval expedition led by Sir John Ross, from a position close to the north magnetic pole in Arctic Canada. The bird duly arrived in Great Britain five days later, having crossed the Arctic and Atlantic Oceans in the frigid autumn weather bearing its message. Flying across the Arctic is not normal pigeon behaviour – breeding by man had adapted a special characteristic. However, in a sense, any migratory bird is pre-adapted as a letter carrier, because it can navigate and return to a known point. Had we spent the past hundred years developing Darwin's work on the selective breeding of pigeons, we might by now have bred an extremely efficient and wholly distinct new species of giant postal pigeon; unfortunately we chose to use jet aeroplanes and the Canadian post office, which combination seems to take two weeks to carry a letter from Canada across the Atlantic.

Species can arise in many different ways. In general the fundamental cause, as Darwin discovered, is natural selection – the fittest survive, the unfit die out. Over time, the organisms or genes that are reproduced are those with some quality that makes them more likely to have offspring that survive to reproduce in the next generation. But there are innumerable wrinkles in this basic process. Sometimes an organism will become isolated (say on an island) and its offspring will be subject to competitive pressures different from the rest of the population: the various different species of finches that Darwin studied on the Galapagos Islands are the classic example of this. Eventually, after many generations, the differences become so great that the isolated group cannot interbreed with the main parental line – a new species appears, or the isolated group may diversify into several distinct species, if there are alternative ways of making a living and no pre-existing competition.

On occasion, a mutation may occur in a member of the species, to produce a very different animal. However,in a sexually reproductive species the mutant can be so different that it cannot interbreed with the rest of the population. If so, it dies out: many mutants simply disappear. If, on the other hand, the mutation is successful, it will be incorporated in the species and, over the generations, the gene responsible will be spread through the whole population. In the Zambesi valley in Africa there is supposed be a small tribe of people, many of whom have only two toes splayed out from the heel; like an ostrich foot, this arrangement may help in walking across sandy country. If this rather startling change were very advantageous, over the centuries people with two toes would become dominant, five-toed people would selectively die out, and – eventually – all people would have two toes. A future palaeontologist would recognize a new species with two toes, not five, taking over in the geological record from an ancestral five-toed species. Sometimes the ancestral line or something like it remains too, while the mutants split off – the worm-like beasts of the Burgess shale may be our ancestors but there are plenty of worms around today too, our cousins.

Why do mutations occur? There are probably many reasons, ranging from natural radioactivity and cosmic rays to simple accidents of reproduction. Although the DNA may change steadily with the generations, with each reproduction the body of each successive organism may show no apparent change from the previous incarnation. Over time, however, only the most important genetic information in the DNA is conserved (because individuals without it die); in particular the nonsense or silent information held in the introns changes steadily. The fossil record of organisms may show no change over long periods and then apparent sudden changes, that may reflect the times when the cumulative effect of small changes finally alters the body of the organism. The body may change suddenly, after many generations of stasis, although the DNA within the cells of the organism has changed slowly over time, generation by generation, as the molecular clock (discussed in Chapter 5) ticks.

Sometimes the unused DNA may come in handy. Imagine a housebuilder, who starts with certain rules of construction, and a plan for a little home (all he can afford). As times passes, he expands the house and adds new rooms for the family, and a new kitchen; the old plans are put away. Then grandma comes to stay. Out come the old construction documents, which are used to build a small home for her alongside the main house. Someone driving by the house once a year would see an apparent sudden change, but the plans are old – they have simply been silent for years and are now expressed. Later, a garage is needed. The old documents can be brought out again, reduced, and used to build the garage. There is an analogous process in life. The molecular biologists call the process 'molecular tinkering'. Organisms probably 'tinker' endlessly with their DNA during the accidents of reproduction. Normally, the silent information in the DNA is simply junk, but a small accidental change can allow an organism to discover new ways to use silent information on rare happy occasions.

There is another way in which change can occur. Sometimes, two organisms may hybridize to produce a distinctive product. Wheat, for instance, is a spectacularly successful plant which was created by two separate genetic events. These events finally produced a plant which now grows in a monoculture across large areas of Australia, Argentina and North America, thousands of miles from the Middle Eastern home of the parental grasses. From the point of view of the wheat genes, this is success indeed. The combination of genes that produced wheat would not normally be viable – the seeds would die – but for the related accident that the wheat plant found a symbiotic partner in humanity. Natural genetic tinkering can occasionally create a new organism by crossing even further over the species barrier. For instance, the plants that can acquire genes derived from bacterial DNA, and the case of the bacterium that has apparently absorbed a gene from its symbiotic host, a fish, have been mentioned already. In these cases, information has crossed not just between species but between kingdoms of life: to go back to a previous discussion the ostrich can, very rarely, learn to swim from the hippos.

FISHES AND PLANTS

Towards the end of the Cambrian, bigger and perhaps better things began to happen, at least from our point of view. Animals with backbones became much more sophisticated.

There is a small living creature today, called *Amphioxus*, that gives some idea of what our distant ancestors looked like. *Amphioxus* is about 5 cm (2 inches) long, and spends most of its life partially buried in sediment in shallow marine water. *Amphioxus* looks roughly like a fish, but is more primitive. Like *Pikaia*, it has a stiffening rod running along its length, not a backbone. It is so small that it breathes by exchange of gases through the general body surface, not with gills. It has few muscles in its gut, and no heart – blood is moved by contractile tissue. The blood has no red cells or haemoglobin. Instead, oxygen is carried in solution in body fluids.

Amphioxus is a very primitive chordate animal. More advanced chordates, with backbones, developed in the late Cambrian and Ordovician. These animals evolved better muscles and more complex nerve structures. The muscles allowed the development of a heart, more active movement of food, and faster gas exchange, which all contributed to the development of a higher rate of metabolism and a more vigorous way of life. The nerve development led to a brain and complex sensory structures, giving the animals good eyes and ears. Bones were probably not present at first, and in the earliest fossil vertebrates bony tissue forms only a cover to the body. This bony tissue may initially have been used to insulate the electrosensory organs present in many primitive vertebrates. Once the ability to form bony tissue had evolved, the animals could then have developed a more general protective cover. Later, by the early Devonian, the ability to make an internal skeleton developed.

In the late Cambrian and Ordovician, the first fish evolved, including jawless creatures very distantly similar to the modern hagfish and lamprey. These fish had, typically, well developed head armour, two eyes (and in some a third or pineal opening), as well as a nostril and a mouth. In modern mammals the third or pineal eye is gone, though the structure still remains as an important organ in our brains; for the rest, this is the plan of the human face. The mouth initially was suited only for mudstraining or filter-feeding, but over time the arches around the gills of the fish evolved into jaws. The gill slits that once existed behind the jaw developed into openings which in us have eventually become part of our ears. Jaws, of course, are better with teeth. Handily, the armour around the heads of the fish was ornamented with ridges of bony material. These were turned into teeth. They were made of a substance known as dentine: the dentine of the armour of the first fish has now, in humans, become the playground of the dentist's drill. During these changes, other modifications happened, too. Small pouches in the gills may have turned into the prototypes both for our lungs and for the swim bladder in modern fish, which helps them orient themselves.

Evolution is a curious process. Because in a sexually reproducing organism, it cannot go backwards, it must accidently tinker with what is available and develop new parts by using whatever lies around in the back yard. A bottom feeder has its eyes above its mouth – we have kept this eminently sensible

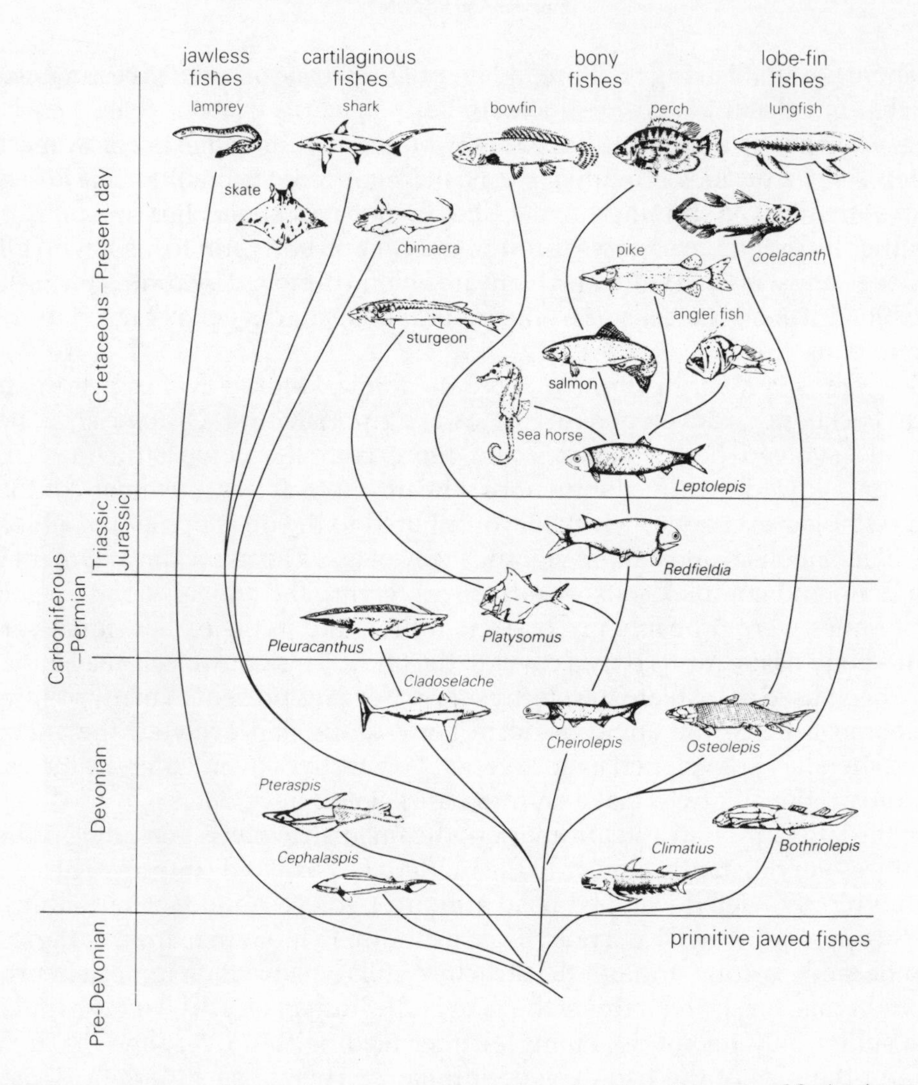

Figure 8.8 The evolution of the fishes (courtesy of Department of Geology, University of Saskatchewan).

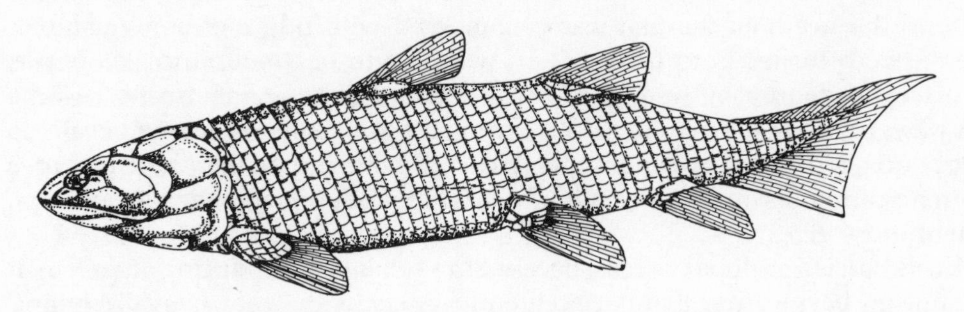

Figure 8.9 *Osteolepis*, an early fleshy-finned fish that lived in the mid-Devonian, about 15 cm (6 inches) long (courtesy of Department of Geology, University of Saskatchewan).

arrangement, but so has the flamingo, for whom it might seem to be a nuisance, as it feeds with its head upside down.

Fish were very successful. The large predators that had existed earlier, the eurypterids (which looked like giant swimming scorpions), slowly lost the struggle eventually and died out; meanwhile the fish diversified and produced fish that ate fish, and fish which ate those. Some were nearly 10 m (30 feet) long. The sharks appeared early on and so did the bony species that were the ancestors of most modern fish.

Figure 8.10 The late Devonian sea. On the right is *Dunkleosteus*, which grew up to 9 m (30 ft) long. Bony armour covered the head and forward part of the body. Bony cutting edges on its jaws served as teeth. The fish on the left is more similar to our own ancestors (painting by Vladimir Krb, Royal Tyrrell Museum of Palaeontology/Alberta Culture and Multiculturalism, Canada).

On land and in fresh water too, evolution was creating distinct ecosystems. One of the interesting contrasts between continental and oceanic environments is that the possibility of geographic isolation is much greater on land (especially on islands), or in a river, than at sea – all sorts of isolated communities can occur. The arthropods – spiders, scorpions and millipedes – and molluscs such as snails and clams were probably the first animals to explore the continents, entering by the rivers, marshes and bogs. Millipede-like animals may have begun crawling on land as long ago as the Ordovician, 450 million years ago.

Some of the oldest evidence for land life comes from Rhynie, near Aberdeen in Scotland. The deposit seems to preserve what was once something analogous to a peat bog, which was suddenly preserved by silicification that replaced the organic structures with SiO_2, perhaps after a volcanic eruption. In the rock are fossils of a whole community of plants, animals and associated fungi. The plants had woody cells and stems. They probably stood in water or wet ground and grew up into the air, much in the way that marsh plants do today. The stems and leaves show pores through which gases were exchanged with the air, and they had a waxy waterproof coating to prevent them from losing too much water to the air and drying up. All field geologists know that swamps are full of bugs, and so was the Rhynie bog. Tiny wingless insects are preserved; they seem to have eaten into the plants.

All plants and animals which live in air have had to evolve a spacesuit. We ourselves are still in a sense marine animals surrounded by a spacesuit, our skin. Our blood is the sea in which our cells live. Freshwater fishes joined the community. As the concentration of various ions in an organism is higher than in freshwater, water will enter a freshwater fish until it dies of dilution or swells

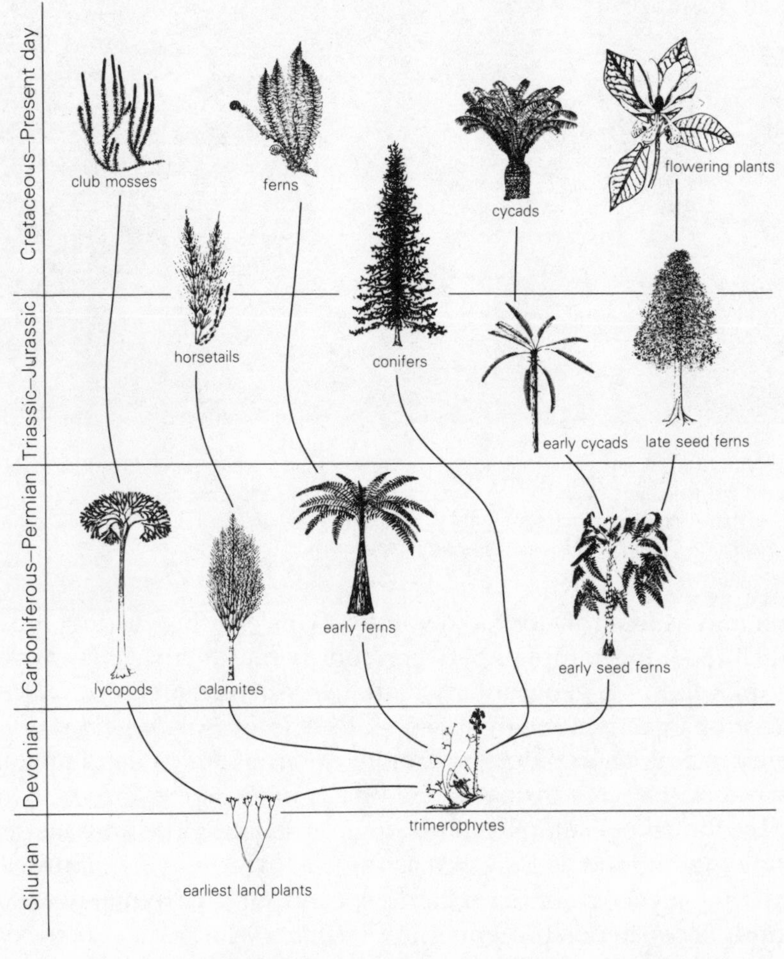

Figure 8.11 The evolution of the plants.

and bursts. The answer was the kidney, which manages the excess water. Snails too had to adapt to problems of this type.

During the Silurian, with roots to draw water out of the soil plants became established on the land itself. These plants had channels made of cells that had died and whose walls had been adapted as conduits for carrying water upwards. The first land-plants were ferns, mosses and horsetails, which release spores to reproduce. The spores must land in a damp place, where fertilization takes place – as a result, these plants could occupy only a very small proportion of the land. Later, more advanced plants evolved mechanisms for creating a tiny pond on the parent plant: in these plants, fertilization takes place when male pollen grains land on a large female spore and are released within it in a drop of water to mate and form a seed. Modern maidenhair trees (ginkgos) and cycads are descended from these organisms. Later, some plants developed with pollen and eggs on the same tree – from these came our pines and firs.

Figure 8.12 Tree ferns, North Island, New Zealand.

By the Carboniferous period forests existed, covering huge areas of land: they have left behind from this period many of the northern hemisphere's coal-beds as a memorial. These were forests of club-mosses, tree ferns and horsetails; there were as yet no flowers or grasses. Tree ferns are surprisingly adaptable – today they grow in settings as different as the cool, seasonally dry, climate in Africa to damp subtropical forest in New Zealand; similarly, cycads (bread plants) flourish from Guam in the Pacific to the dry Karoo semi-desert and grasslands of South Africa. In one place in South Africa a cycad forest exists today – a vision from the past set up by a local tribal cult. These plant families may once have colonized much of the Earth.

Plants are the food of animals. Here was a food source to enable an animal community to spread out across the land. Equally important, plants, especially land plants, act in concert with animals to process the atmosphere. Animals

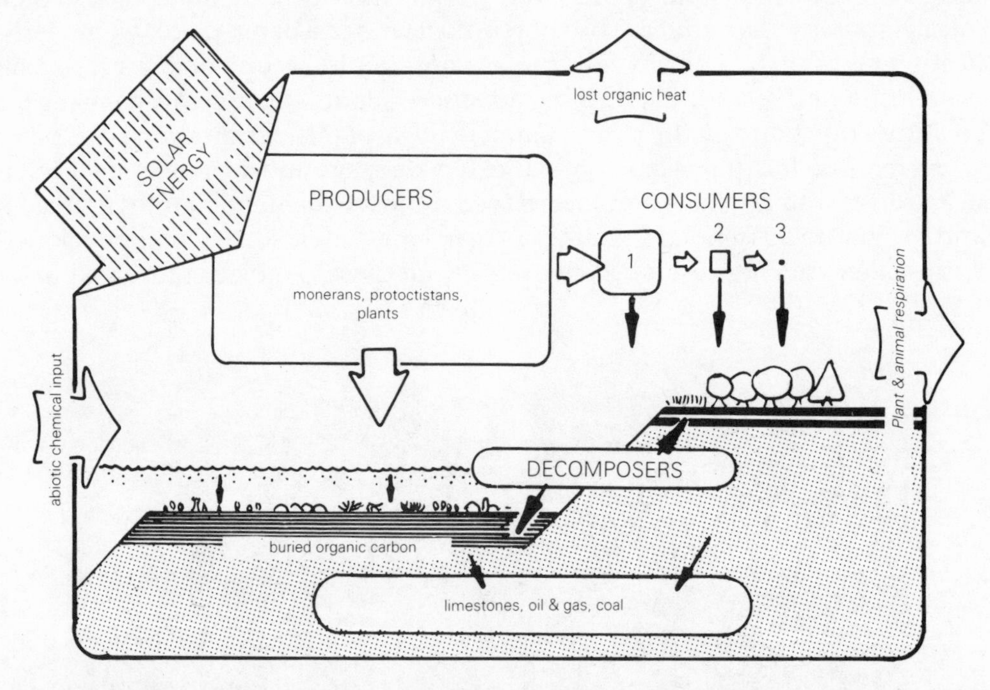

Figure 8.13 Energy flows in the biosphere. Solar energy flows via producers and consumers, most to be lost as organic heat, some to be stored by burial (from Copper, P. 1988. *Geoscience Canada* **15**, 199–208).

need oxygen, plants need carbon dioxide. In the oceans, plant plankton release oxygen. Oxygen is also mixed from the air into the water by dissolving across the air/water boundary, especially with the help of bubbles produced in the breaking whitecaps of oceanic waves. On land, the interaction between plants and the air is much more direct. Land plants can hold up dense structures of leaves, overlapping to catch light, and with large surface areas exposed to the atmosphere. Leaves act as chemical factories, processing the atmosphere. The level of carbon dioxide in the air is determined today by the biological turnover, especially on land: very generally speaking, if there is abundant carbon dioxide, plants will flourish and remove it; if there is little CO_2, there will be less growth until decay and volcanic degassing have replenished the supply.

Atmospheric oxygen is managed over a longer timescale. Fire is a chief instrument in managing oxygen, and this management may have evolved as early as the Carboniferous, when plants began to escape the swamps and

(a) Grazing–browsing food chain

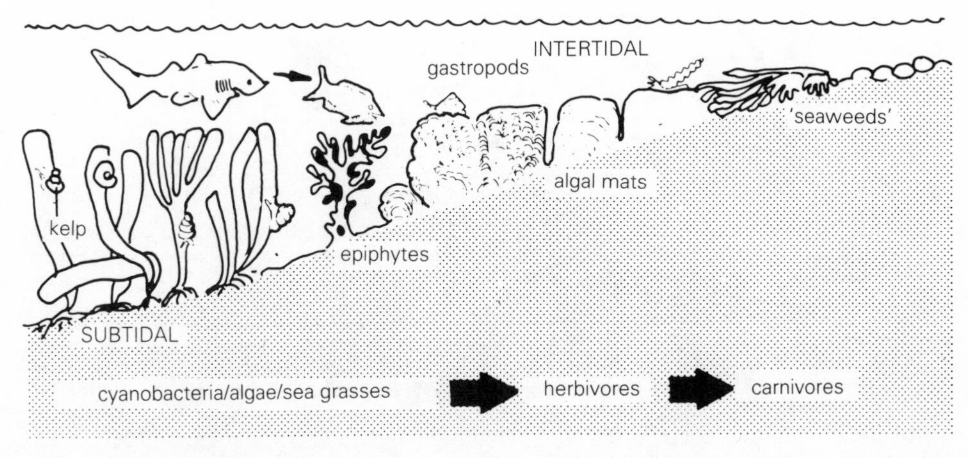

Figure 8.14a The grazing–browsing food chain in the seas. Algal or bacterial mats are grazed, and 'leafy' algal growths and sea plants are browsed by herbivores such as snails and fish. Such food chains typically occur in very shallow water, and began at the beginning of the Cambrian, when stromatolitic algal mats, in particular, were reduced in abundance by animal grazing (from Copper, P. 1988. *Geoscience Canada* **15**, 199–208).

(b) Suspension feeding food chain

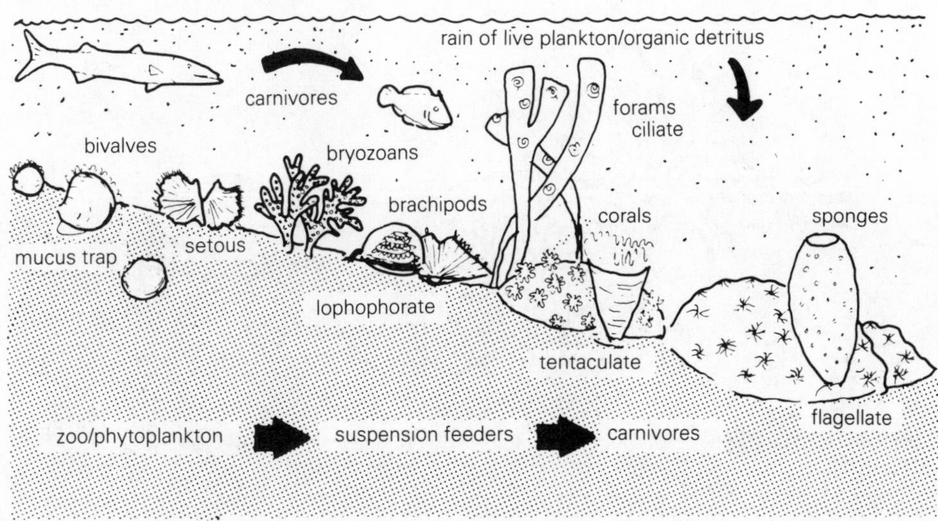

Figure 8.14b The suspension feeding food chain. Palaeozoic suspension feeders were typically bottom-living invertebrates, such as corals or brachiopods. They lived on plankton and dead organic matter in the water (from Copper, P. 1988. *Geoscience Canada* **15**, 199–208).

(c) **Detritus feeding food chain**

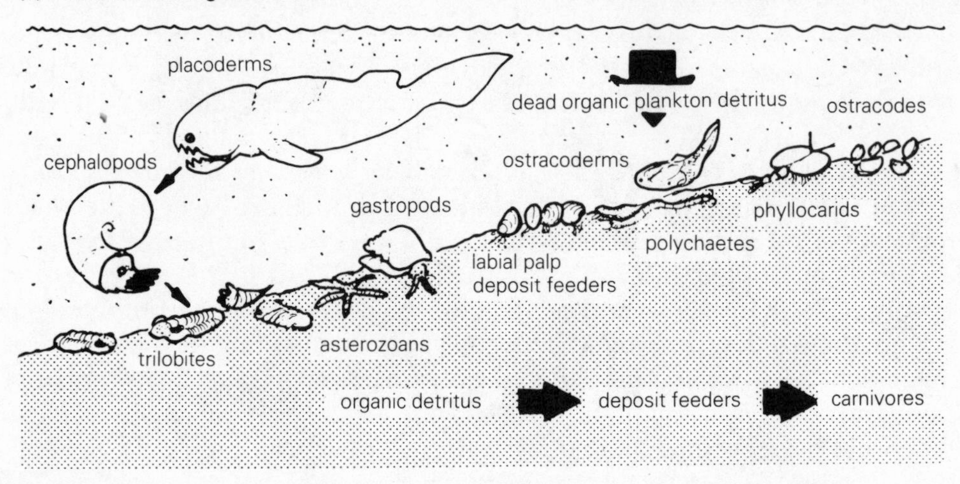

Figure 8.14c The detritus feeding food chain. In the Palaeozoic, debris, such as plankton detritus, was eaten by jawless fishes and various deposit feeders. Animals such as cephalopods ate the deposit feeders, and were themselves eaten by bigger carnivores (from Copper, P. 1988. *Geoscience Canada* **15**, 199–208).

(d) **Deep sea chemotrophic food chain**

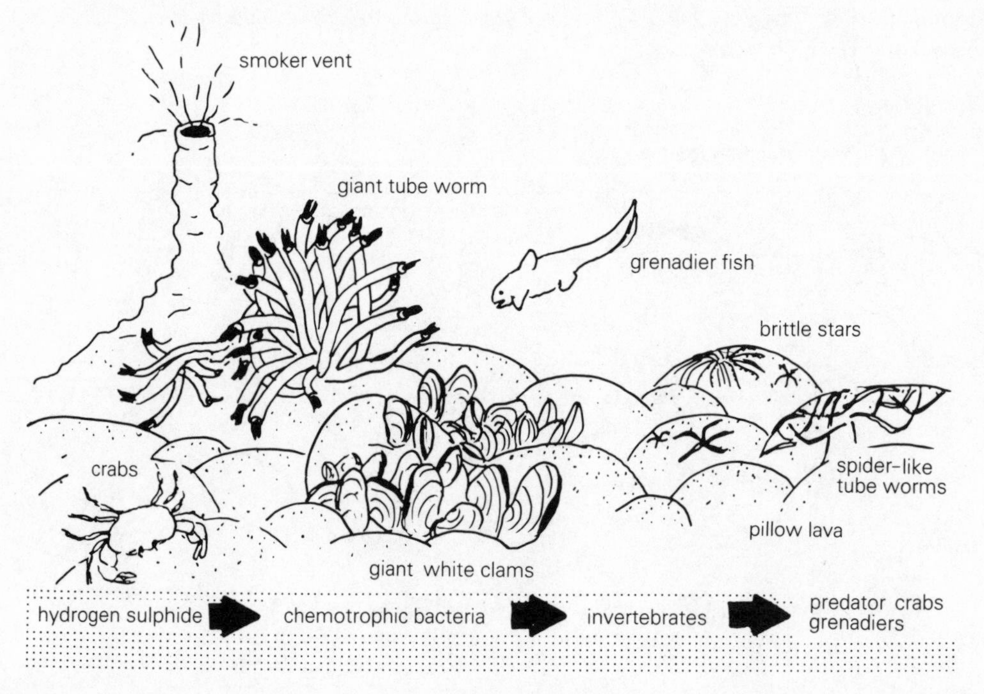

Figure 8.14d Deep sea chemolithotrophic food chain around hydrothermal vents near volcanoes and on mid-ocean ridges. Bacterial growth around vents supports a variety of invertebrate animals, in a food chain. The system is not wholly detached from the photosynthetic system in the rest of the sea and on land, because it needs a supply of oxidized chemicals in seawater, supplied via photosynthesis (from Copper P. 1988. *Geoscience Canada* **15**, 199–208).

colonize the dry land. J. Lovelock and his colleague, A. Watson, have suggested that today one of the controls on the oxygen content of the air is the burning of the northern forests and the tropical dry bush. These fires may help to set the oxygen level at just below the point when everything burns if lightning strikes. Fire is used as a competitive weapon by plants. Plants which have seeds that can survive intense fire tend to have leaves and stems that burn intensely, destroying surrounding plants. After the fire, when the seeds sprout, the surrounding species have been wiped out. Many pine trees, for instance, have adopted this strategy. Other plants survive by resisting fire, and not burning. In many forest and scrub lands the net effect of the competition is that a fire ecology has developed between species competing for survival. Fire management was initially probably of only local importance. Slowly, as dry regions and temperate and high latitudes were vegetated, the oxygen management would have become more effective.

FURTHER READING

Brasier, M.D. 1980. Microfossils. London: Allen & Unwin.
Margulis, L. & K.V. Schwartz 1988. *Five kingdoms: an illustrated guide to the phyla of life on Earth.* New York: W.H. Freeman.
Stanley, S.M. 1989. *Earth and life through time,* 2nd edn. New York: W.M. Freeman.
Whittington, H.B. 1985. *The Burgess shale.* Yale: Yale University Press.

9
The changing land

THE DANCE OF THE CONTINENTS

Geologists have a reputation for struggling through swamps, fending off moose and elephant, walking thirstily across deserts, or descending into the mouths of volcanoes, but some of the most courageous of all fieldwork was in the most tragic of circumstances, during the exploration of Antarctica. There is much geological evidence that parts of Africa and India have once been under ice, while at other times, thick forests have flourished in what is now Antarctica. A critical part of the story was collected by Scott's expedition on their doomed race against Amundsen to the South Pole. On their way, Scott's party stopped to collect rock samples, and they refused to jettison the heavy stones from their sledge even when they knew they were dying; the expedition was primarily for science, then for publicity. When their bodies were found, the samples and notebooks were still with them.

One of Scott's precious samples was a specimen of a fossil plant, *Glossopteris*. It proved to be a critical piece of evidence in the proof of continental drift, a decisive piece in the jigsaw puzzle of fitting together the southern continents.

The scientific reassembly of the continents began earlier. In early Victorian times, the young Charles Darwin communicated to the Geological Society an interesting note from a traveller, Andrew Geddes Bain, who wrote about the horns of a huge ox preserved in the interior of South Africa. Perhaps the contact continued, we do not know, but some years later Bain was working on the rocks and fossils of the Cape Province of South Africa and noted their extraordinary similarity to the material collected by Darwin in the Falkland Islands during the voyage of HMS *Beagle*. There must have been a connection, reasoned Bain. Other rocks also implied that a great land mass had once existed to the south of South Africa. After Captain Cook and Charles Darwin, humanity was able to have a world view: scientists could see across oceans.

Science is global, and scientists, especially earth scientists, form a worldwide community. Few other branches of society have so clear a view of the Earth as a whole; most professions see only their local environment, or their own nation. The scientific community today, as then, is small and transnational. Of the scientists, it is the geologists and the naturalists who see the most of the Earth. In the nineteenth century, this small circle looked at the globe with curiosity. Today, the curiosity is still there, but it is associated with deep concern about the state of our planet.

One of those who followed Bain in the tradition of Southern African geology was Alex du Toit, who in the 1930s managed successfully to put all the pieces together. In his book *Our wandering continents*, he reconstructed a former great continent of Gondwanaland, comprising what is now South America, Africa, Antarctica, India, Australia, New Guinea and New Zealand. Du Toit realized that there was overwhelming evidence – from the distribution of modern plants and animals, from fossil plants such as *Glossopteris*, from animal fossils, from comparison of rock types, from the record of a vast southern hemisphere glaciation which even extended to India, and from the shapes of the continents – that all these modern land masses had once been grouped together as a huge supercontinent. For it, he adopted the name *Gondwanaland* given by a nineteenth-century geologist, Suess, who also gave us the term 'biosphere'. The name Gondwanaland is tautologous,but used by most geologists.

Our wandering continents is one of the most important of the geological books of the twentieth century, although it was initially rejected by the scientific community. Together with the work of Arthur Holmes in Britain, it laid a firm factual base to the notion of continental drift, the shuffling dance of the continents. This idea had earlier been proposed by Taylor in the USA and Wegener in Europe (to whom du Toit's book is dedicated). Later, the idea of continental drift was supported by evidence from rock magnetization that showed that the continents had moved across latitudes. This magnetic research was done in the 1950s and 1960s, especially in Cambridge and Newcastle, with much of the Gondwana work being done in Africa, in Harare. Computer models of the fit of the Atlantic continents further supported the idea. Continental drift was eventually explained in the larger framework of plate tectonics (Chapter 1).

For half a century, the notion of continental drift, which we now know to be true, was opposed by the geological establishment. There is an interesting moral to this saga. It is not that established science denied the validity of a good theory; that is correct, but unfair. A deeper observation is that the structure of science is built so that wrong ideas are filtered out by rigorous criticism. This means, of course, that good ideas too are frequently rejected, and it is often very difficult for a valid idea, such as continental drift, to be accepted, especially when the details of the idea can be shown to be wrong, as was the case with Wegener's early notions about drift. This is part of the price of scientific rigour. The arguments *against* drift were powerful and stated intelligently by careful scientists of integrity; the arguments *in favour* were initially highly speculative and weakly supported by fact. It was the need to counter the arguments against drift that shaped du Toit's book and drove the later search for geophysical evidence; scientific dispute is essential to scientific progress. In consequence of the debate, it took fifty years for the theory of drift to be accepted. But if, on the other hand, we were to accept speculative ideas immediately, we would inevitably accept much nonsense. Our hard-won firm ground of science would turn to a shifting quicksand of notion and fantasy. What is important is that we allow sufficient freedom in our structure for strongly held dissent. Continental drift survived by retreating from the scientific centre to the fringes, where it was supported by courageous advocates in the British north, South Africa and Tasmania. Here there was freedom to dissent. It is also important that we strive not to attack the opposing side unjustly: in many scientific debates the heretics are unfairly branded as lunatics.

Continental drift always evokes images of Gondwanaland breaking up, or of the Atlantic opening, but the drunkard's walk (or is it a dance?) of the continents is literally as old as the hills which it created. Ever since the first landmasses were created, continents have been moving slowly across the surface of the planet at velocities of a few centimetres (or inches) per year, at times aggregating into huge supercontinents, at other times dissembling into fragments.

Our certain knowledge of how the continents have moved goes back only as far as the record of the spreading of the ocean floor, which is really useful for a little more than a hundred million years. Before that, we must guess from fragmentary evidence, which includes the knowledge from magnetic evidence of the latitude (but not the longitude) in which rocks were laid down, and the general geological good sense that was the basis of du Toit's reconstruction of Gondwanaland.

In the Carboniferous period, the world was probably a planet with a vast ocean and spread-eagled giant continents. In the south was the supercontinent of Gondwana, comprising modern South America, Africa, India, Antarctica and Australasia. Europe probably lay across what is now the northwestern bulge of South America (Panama sticking into Greece, London on the Equator) joined to North America to the west (with Winnipeg and San Diego also on the Equator). Siberia, China and mid Asia may have formed a northern continent, perhaps separated from modern Russia by an Ural sea. During the Permian, roughly fifty million years later, the core of Gondwanaland (Antarctica) lay near the South Pole, with India not far off. A Permo-Carboniferous glaciation spread across Gondwana; the difficulty of explaining how a southern glaciation could reach across the Equator to cover India in its modern position was one of the decisive pieces of evidence behind du Toit's advocacy of continental drift. India was, in fact, close to the South Pole.

In North America this time is marked by a major break in the rock record, which perhaps was a consequence of the erosion that occurred when the shallow sedimentary basins on the continental surface were drained in a period of lowered sea-level during the glaciation. The Earth today is in a similar glacial

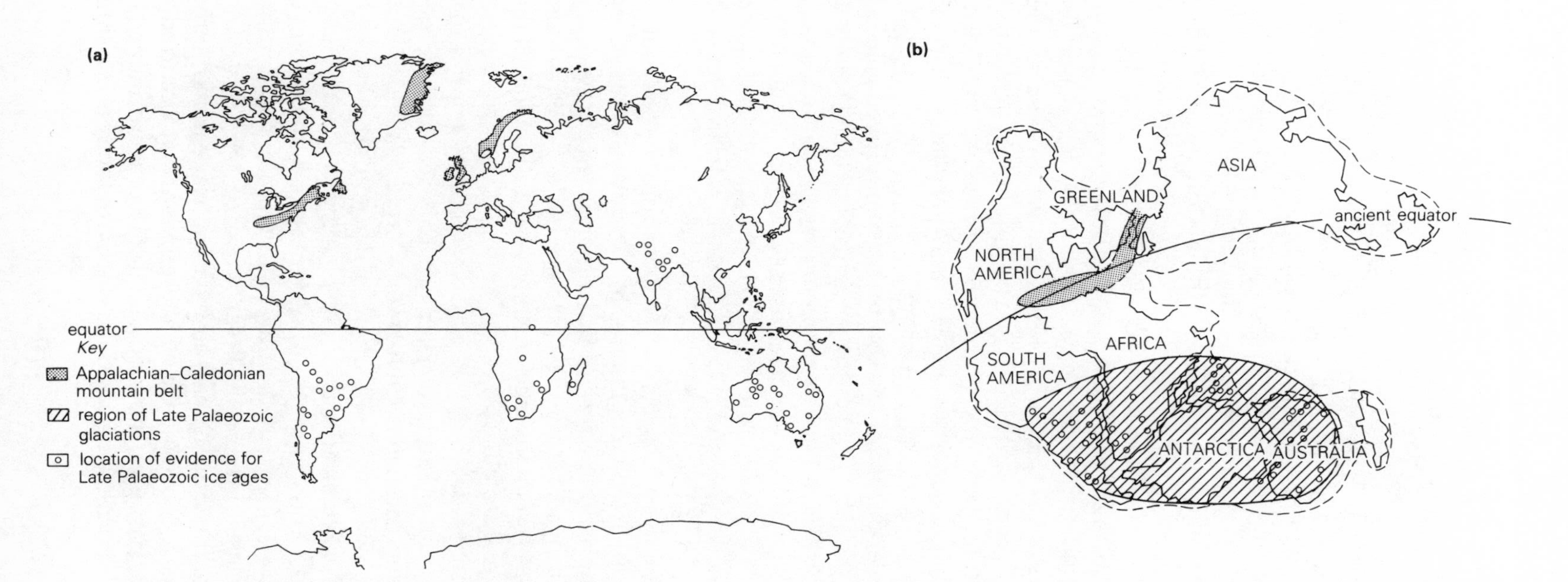

Figure 9.1 The reconstruction of Pangaea, a brief-lived super-continent. (a) The fragments today. (b) A possible reconstruction of Pangaea 200 million years ago (courtesy of Department of Geology, University of Saskatchewan).

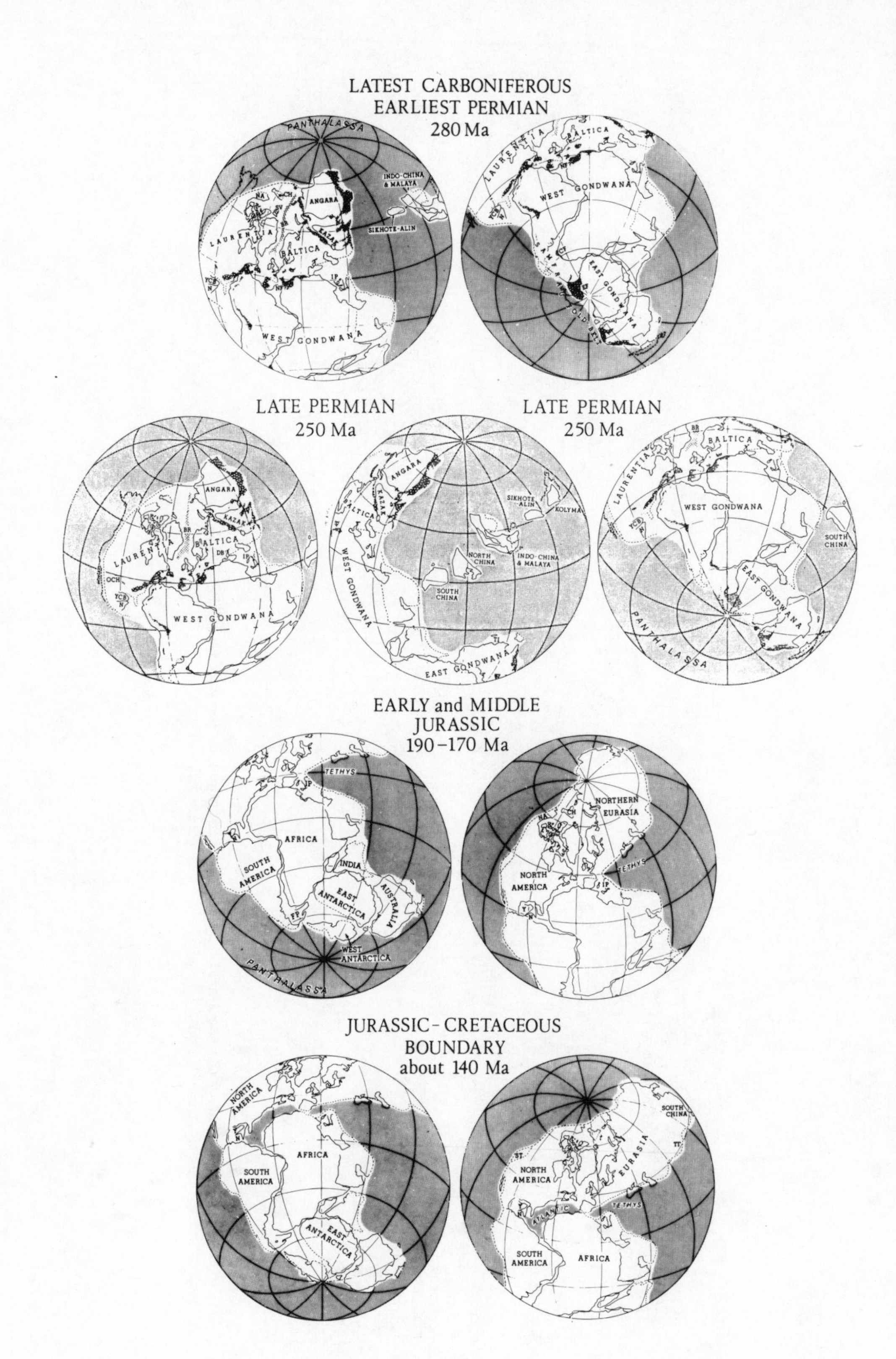

LATEST CARBONIFEROUS
EARLIEST PERMIAN
280 Ma
PANTHALASSA
INDO-CHINA & MALAYA
ANGARA
SIKHOTE-ALIN
LAURENTIA
BALTICA
KAZAK
WEST GONDWANA
EAST GONDWANA
LATE PERMIAN
250 Ma
LATE PERMIAN
250 Ma
ANGARA
KAZAK
BALTICA
LAURENTIA
WEST GONDWANA
SIKHOTE ALIN
KOLYMA
NORTH CHINA
INDO-CHINA & MALAYA
SOUTH CHINA
EAST GONDWANA
PANTHALASSA
EARLY and MIDDLE
JURASSIC
190–170 Ma
TETHYS
AFRICA
SOUTH AMERICA
INDIA
EAST ANTARCTICA
AUSTRALIA
WEST ANTARCTICA
PANTHALASSA
NORTHERN EURASIA
NORTH AMERICA
JURASSIC - CRETACEOUS
BOUNDARY
about 140 Ma
NORTH AMERICA
AFRICA
SOUTH AMERICA
EAST ANTARCTICA
SOUTH CHINA
EURASIA
ATLANTIC
TETHYS

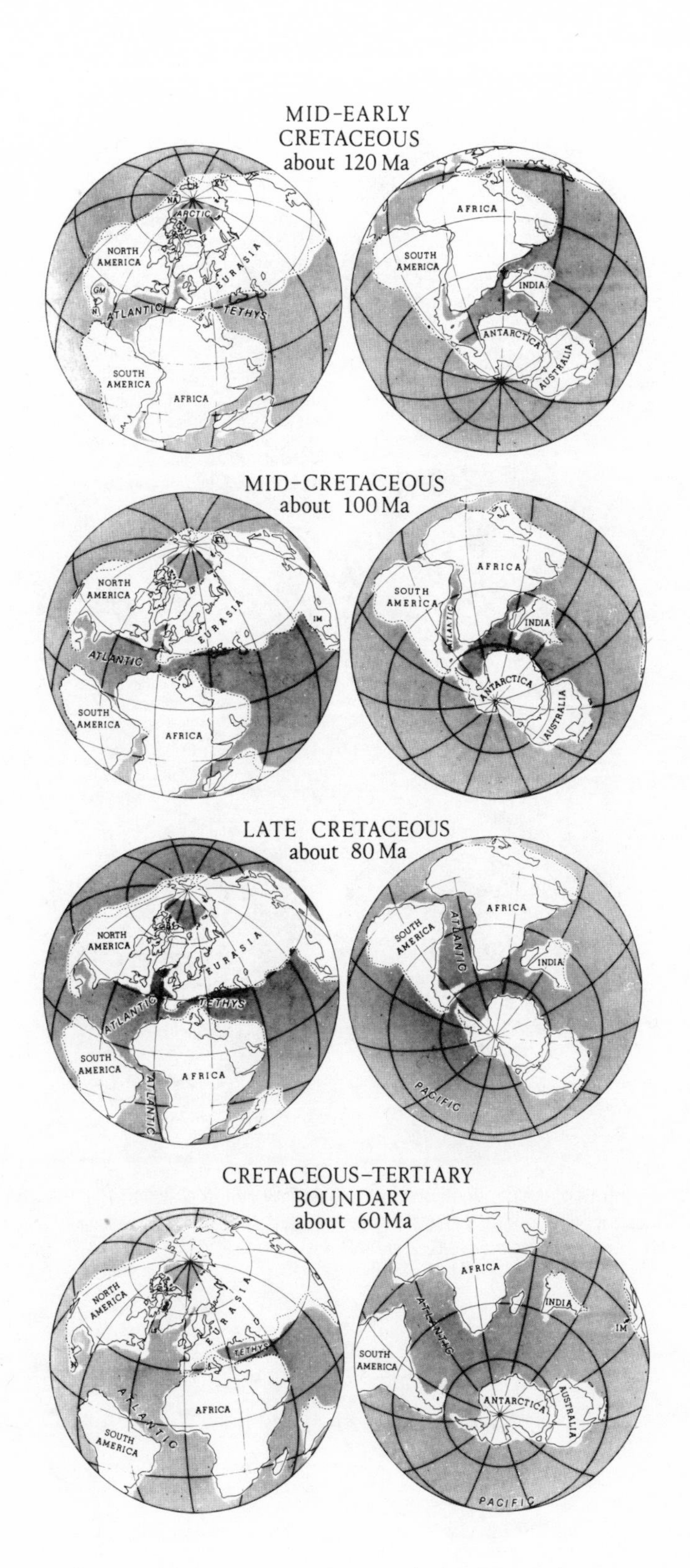
MID-EARLY
CRETACEOUS
about 120 Ma
ARCTIC
NORTH
AMERICA
EURASIA
ATLANTIC
TETHYS
SOUTH
AMERICA
AFRICA
AFRICA
SOUTH
AMERICA
INDIA
ANTARCTICA
AUSTRALIA
MID-CRETACEOUS
about 100 Ma
NORTH
AMERICA
EURASIA
IM
ATLANTIC
SOUTH
AMERICA
AFRICA
AFRICA
SOUTH
AMERICA
INDIA
ANTARCTICA
AUSTRALIA
LATE CRETACEOUS
about 80 Ma
NORTH
AMERICA
EURASIA
ATLANTIC
TETHYS
SOUTH
AMERICA
AFRICA
ATLANTIC
AFRICA
SOUTH
AMERICA
ATLANTIC
INDIA
PACIFIC
CRETACEOUS-TERTIARY
BOUNDARY
about 60 Ma
NORTH
AMERICA
EURASIA
TETHYS
ATLANTIC
AFRICA
SOUTH
AMERICA
AFRICA
INDIA
IM
ATLANTIC
SOUTH
AMERICA
ANTARCTICA
AUSTRALIA
PACIFIC

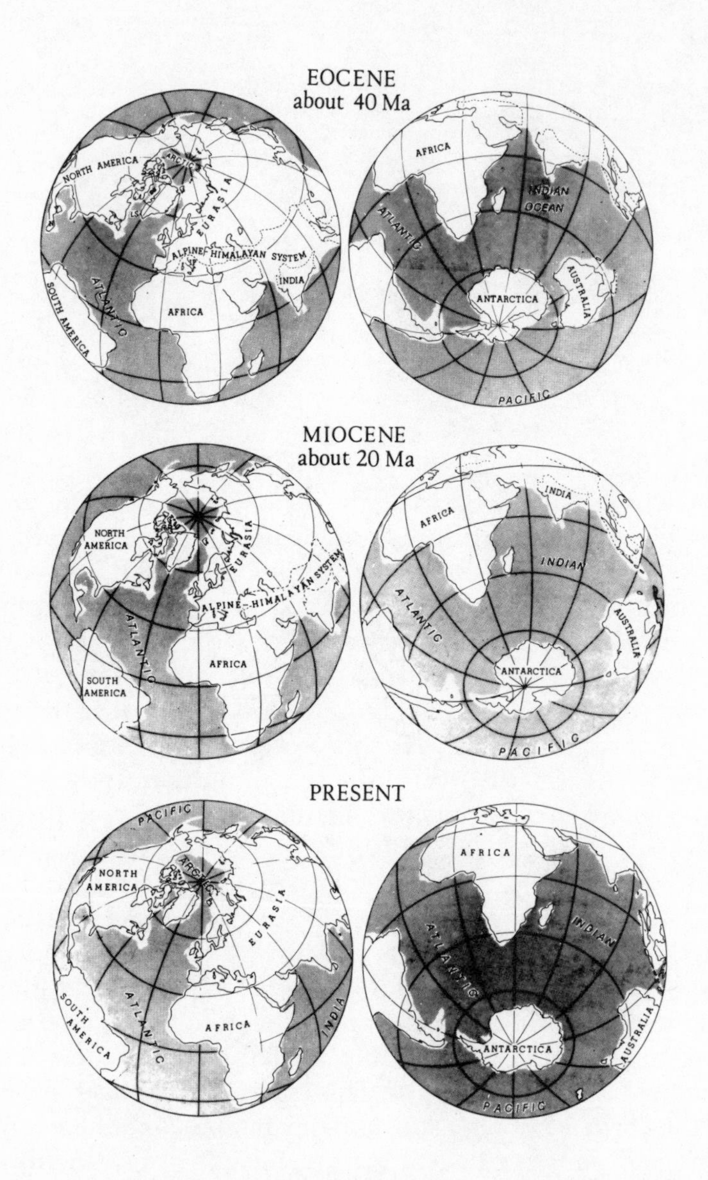

Figure 9.2 The movement of the continents over the past few hundred million years. These reconstructions are by E. Irving. Other reconstructions differ slightly from this, but the main reassembly is well established (from Irving, E. 1983. *Geophysical Surveys* **5**, 299–333).

episode – so much water is locked up in the Antarctic and Greenland ice-caps that the wide low continental plains are exposed. Eventually, if the ice all melts, London, New Orleans and Calcutta will go back to their normal submarine state.

THE MECHANISM OF CONTINENTAL DRIFT

Continental drift is a consequence of plate tectonics, the process by which heat is transferred out of the Earth and eventually radiated to space (though the Earth's heat loss from the interior is minute compared to the heat received from the Sun and then re-radiated). A boiling pot of oatmeal porridge was considered in Chapter 1 as a good analogue of plate tectonics. The surface has lines of upwelling where hot porridge is coming up, areas where porridge is moving across the surface, and narrow zones where porridge (now colder) falls down again into the inner depths of the pot. Now imagine putting a ping-pong ball on top of the porridge. It would bob around but would not go down. Continents are not ping-pong balls, and the analogy breaks down if it is taken too far, but they are, in a sense, rafts that float on the surface of the Earth's interior.

Continents are mainly passive in this process. Once they have been created – as the scum which rises up from the remelting that goes on above regions where water is carried down the subduction zones – they simply ride across the surface. On the modern Earth, the Atlantic is widening by a couple of centimetres (an inch) per year, and to accommodate this the floor of the Pacific Ocean is descending below the western flank of the Americas and the margins of Asia and New Zealand. As far as we know, the Earth is not expanding. This idea of expansion was once a popular and reasonable hypothesis to explain continental drift, but, in order to expand markedly, the interior of the Earth would have to be heated considerably, perhaps even to the extent that a large part of it melted or changed to gas. This is not taking place. We know that the mantle of the Earth is not now substantially molten, and there is no heat source to perform the expansion. Instead, our planet has had a roughly steady radius over time, though it may have contracted slightly since the Archaean.

When continents open apart, the creation of new plate surface in one region is compensated by closure and destruction of plate surface elsewhere. The surface created or destroyed is oceanic, so the effect is that the continents ride in the plates. However, the picture is not simple – subduction takes place, commonly, next to continents or continental fragments, and splitting-apart often begins within continents, as in modern East Africa and Ethiopia. Moreover, on occasion continents are indeed subducted, though apparently not deeply. Continents are involved in plate tectonics as active participants, not just as passive passengers. India has under-ridden Tibet, so that there is a double thickness of continental crust under the Himalayas. Eventually, this will be eroded and deep-level rocks will be exposed on the surface: the continent will return to a normal thickness. Africa is beginning to descend under the southern coast of Crete: this geological accident and the steep harbourless coast it has produced had major consequences for St Paul in his voyage to Rome.

Supercontinents may have large-scale effects on the inner movement of the Earth. It is possible that the aggregation of a supercontinent has an effect on the internal distribution of heat in the mantle, because continents act as blankets over the mantle. Possibly, the Earth's history has been marked by episodes in which all the continental material collided together to form a supercontinent, followed by the build-up of heat, and then rifting, after which fragments of continent skittered across the globe, eventually to bump into each other and re-aggregate several hundreds of millions of years later. We are in such a phase of scattering and distribution today.

As the continents move, their plants and animals ride with them, sailing in huge Arks of life. Within each Ark, a distinct ecosystem evolves. New Zealand is an example. Darwin commented that the 'endemic productions of New Zealand . . . are perfect one compared with another'. Yet the products of another Ark may be more aggressive or efficient, so we see the New Zealand life, as Darwin put it, 'now rapidly yielding before the advancing legions of plants and animals introduced from Europe'. The distribution of life is thus dependent on the geographical consequences of plate tectonics. Climate, too, depends on the accidents of drift – especially when a large continental mass gathers over a pole and becomes one of the factors in triggering glaciation.

The story of continental drift is written in the distribution patterns of life. An illustrative example is the distribution of modern freshwater fish. For instance, the lungfish and the bonytongue fishes are mostly restricted to the fragments of Gondwanaland (although some bonytongues have invaded Indonesia). The carp-like fishes occupy the northern continents but have invaded Africa, though not South America, Australia or Madagascar. A group of fishes known as the characins, which includes piranhas, live only in South America and Africa, reflecting the late separation of these two fragments of Gondwana.

One of the most interesting fish groups is the cichlids, which occur in South America, Africa, Madagascar and India, all of which were pieces of Gondwana. Recently, however, *Tilapia mossambica*, an African cichlid, has been spread by humanity to Indonesia, India, the Philippines, Taiwan, Trinidad and Texas. Tilapias are excellent eating fish, but one wonders at the wisdom of colonizing the world with them. Lake Victoria used to have a complex ecology including over 170 cichlid fish species, which provided a delicious diet for the local people. Development came and the predatory Nile perch, Tilapia and other species were introduced: the fragile equilibrium collapsed and today we have a depopulated lake with few species. Humanity is now considering doing the same to Lake Malawi. A similar wave of extinction took place in the North

American Great Lakes following the opening of the St Lawrence Seaway. Of course, introductions have produced successes too, such as the lovely trout streams of Zimbabwe's Eastern Highlands, or the introduced fish of New Zealand's lakes, but we run great risks when we randomly scatter genes around the planet, whether it be Nile perch to Lake Victoria or salmon to Chile.

Fish have occupied the seas and the fresh waters, but they also managed to colonize the land. Our own ancestors were fish, sometime in the Devonian.

THE ARRIVAL OF THE LAND ANIMALS

A few modern fish have lungs. Lungs seem not to be newly evolved gadgets, but relicts of very, very ancient characteristics. Modern fish with lungs include the lungfish, which have fleshy fins, and a few ray-finned fish. They occur on fragments of Gondwanaland. The lungs are used in periods of seasonal drought when rivers and pools dry up and the fish need to obtain oxygen directly from the air.

Lungs were characteristic of primitive fish. We do not know why. Just possibly when they first spread widely there was less oxygen in the air than today, so that less oxygen was mixed into the water by wave-formed bubbles (or, perhaps, the Earth was much less windy, so that the air mixed poorly into the water). If so, the first fish may have needed to gulp in air from the atmosphere to supplement what they could extract from the water with their gills. Or perhaps lungs were simply, by evolutionary accident, present in the first fish and then were slowly done away with in favour of gills.

Whatever the reason, the lungs of early fish pre-adapted them for life on land. Much of evolution consists of some characteristic being put to a use quite different from its original purpose, just as the wartime German V-2 rocket evolved into the US and Russian space vehicles. Florida fishermen, in quiet moments between space rocket launches, can tell tall tales of walking catfish that do occasionally leave the water for brief periods. At some distant point in the Devonian, possibly on mud-flats or brackish deltas, some fleshy-finned fish similarly took to spending part of the time on land. These fish had lungs, strong fins, and were to some extent pre-adapted to the land. A need to wiggle from pond to pond would soon have eliminated those with weak fins and encouraged the survival of the strong. The descendants of these early land-fish became the four-legged animals, including the amphibians, the same broad class of animals that includes modern frogs and which are still clearly linked to water.

The ancestral fish have been extinct for hundreds of millions of years, and for many years lecturers in vertebrate palaeontology declared that there were no living equivalents except, distantly, the lungfish. However, in 1938 a fisherman from East London, South Africa, caught an unusual 1.5 m (5 foot) long fish in

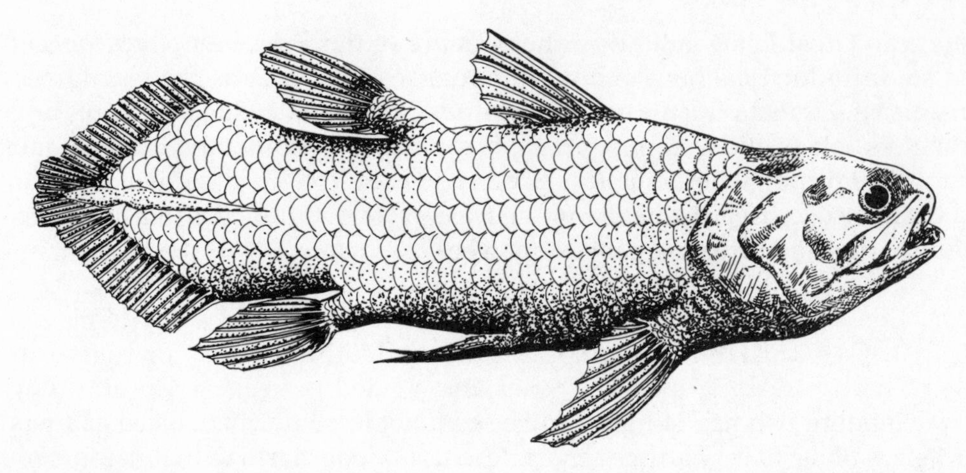

Figure 9.3 The Coelacanth. The fish is about 1.5 m (5 ft) long (courtesy of Department of Geology, University of Saskatchewan).

deep water. It was inspected by Miss Latimer, of the local museum, who realized how special it was and told Dr J. Smith, the fish expert of Rhodes University. Smith immediately recognized it as a coelacanth, a type of fish similar to our distant ancestors. While some of the descendants of the early

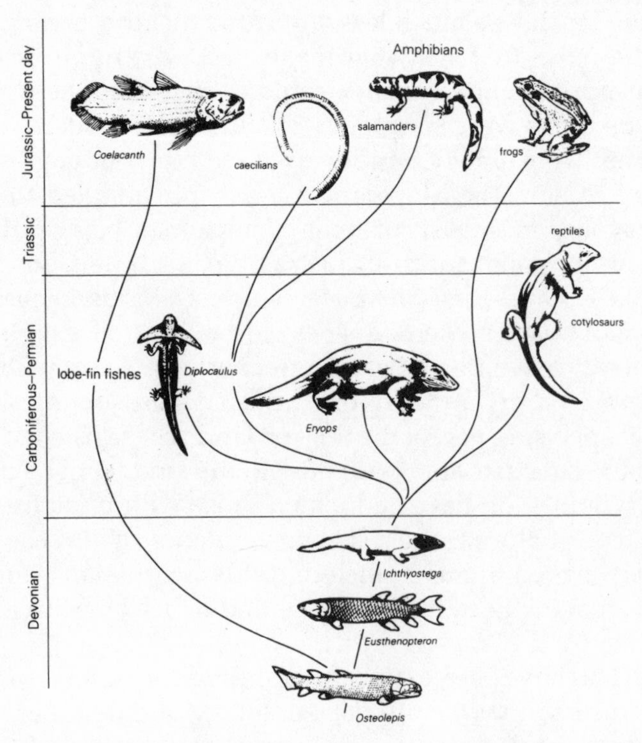

Figure 9.4 The evolution of the amphibians (courtesy of Department of Geology, University of Saskatchewan).

coelacanths took to the land, the others seem to have retreated to the deep salt seas, leaving no fossil record and thought to be extinct until suddenly reappearing in a fisherman's net. Recently the fish has been filmed in its natural setting, behaving more like a gymnast than a fish, with unusual twists and turns. The animal is remarkably like reconstructions of our ancestors. Even its ears are closely similar to ours. It also bears live young, though this was probably not a feature of our ancestors.

In the Devonian the first amphibians appeared. The early amphibians were clumsy-looking animals, with large skulls, short trunks and stocky, splayed legs, but also with lungs and the ability to live – for the most part – on land, or at least in damp swampy places. Their skins needed to keep moist, and they returned to the swamps to breed, laying eggs and sperm on the water as frogs do, and having fish-like, gill-breathing tadpoles or larvae. One of the best examples known of this early amphibian community is preserved in lower Carboniferous rocks from East Kirkton, in Lothian, Scotland. The rocks are about 340 million years old. The host rock is a limestone laid down in shallow pools close to hot springs. The land community had a wide variety of life – amphibians, and also millipedes, scorpions, eurypterids (large arthropods, looking slightly like gigantic scorpions), harvestmen and much plant material.

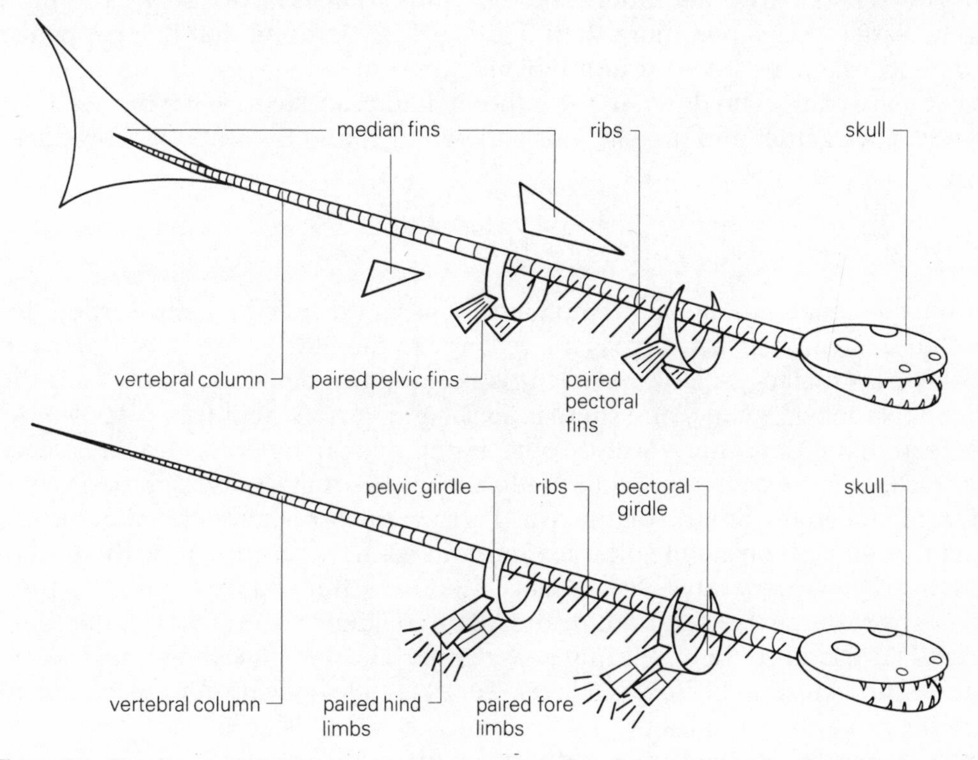

Figure 9.5 The characteristic features of vertebrate animals. Top: fishes; bottom; four-footed land animals (courtesy of Department of Geology, University of Saskatchewan).

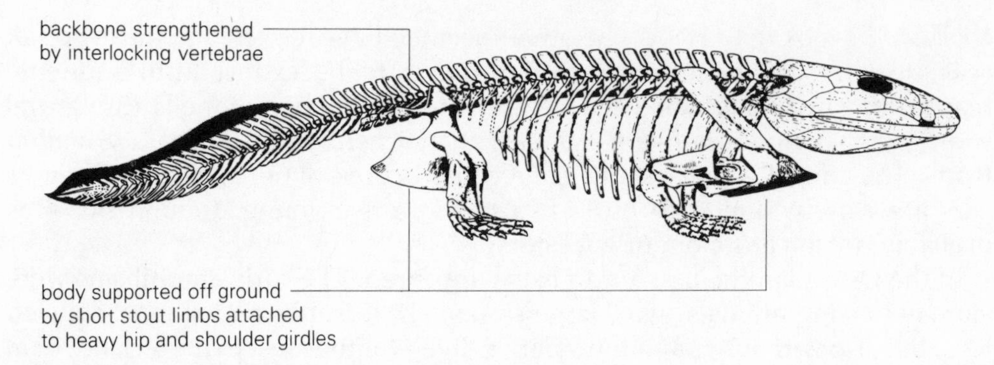

Figure 9.6 The skeleton of *Ichthyostega*, which lived in the Late Devonian. About 1 m (3 ft) long (courtesy of Department of Geology, University of Saskatchewan).

The early amphibians may have fed in the water, or eaten fish and other organisms stranded at the edges of lakes and swamps. The larger land invertebrates may also have been a nourishing food supply.

The amphibians, from their fish ancestors, had evolved the basic body style of spine, head, four limbs and five fingers and toes. All four-footed life, birds and ourselves follow this pattern, though some animals have lost parts of it and some early species had more than five digits. Everything that has happened among vertebrates since the amphibians came out of the water is rooted in the basic fishy plan, even down to the bones in hands and feet. What has varied is the way we gather and manage our body energy, and the way we reproduce.

EGGS

Animals sometimes behave strangely in order to breed. Some turtles, for instance, swim half-way across the Atlantic to the little island of St Helena in order to lay their eggs. Now St Helena is a beautiful and historic place, being the site of Napoleon's exile, and there is a vast and very ancient (pre-Napoleonic) tortoise in the Governor's tennis court, but it is not an obvious pleasure palace for turtles. The reason most commonly advanced for this odd migration is that it began when the South Atlantic was narrow: the turtle ancestors that took to laying their eggs on small volcanic islands in the new ocean, where there were relatively few predatory land animals, stood a better chance of passing their genes on to the next generation. As the Atlantic widened, the turtles shifted from island to island as new volcanoes appeared and old volcanoes died, were eroded and subsided below sea level. Today, St Helena and Ascension Island are the modern examples of this continuing series of islands and remain safe places, guarded by the British authorities, for turtle eggs.

Coelacanths give birth to live young, but most fish do not. The first animals to colonize the land seem not to have found it advantageous to waddle around on land carrying their young internally. Eggs were preferred, and it may have been because of egg-laying that the land was first occupied. For our ancestors, laying eggs on land may have developed as a very ancient stratagem to reproduce safely. In the late Palaeozoic, the land may have been a secure place to escape marine predators, as it was populated only by insects, arachnids and snails. Sophisticated eggs, which could be laid on land (unlike frog eggs) were duly developed. Like a closed spaceship, these eggs contain a life-support system.They are covered by a protective shell and membranes that permit gas exchange into the sea of fluid they enclose, the amnion, in which the embryo can develop and reach a relatively advanced stage before it is born. From this, they are called **amniotic** eggs. Sometimes eggs are retained within the mother, but in general they are laid, and laying an egg well away from water can be a very effective strategy for reproduction. Penguins today in New Zealand do exactly this – they can climb steep 150 m (500 ft) hillsides to lay their eggs safely away from predators (before cats and weasels were introduced).

THE EARLY REPTILES

The East Kirkton limestone in Scotland, which preserves such a diverse community of life, also contains the record of the arrival of animals that laid amniotic eggs. This is the fossil of an amniote, an animal that was one of the group that includes reptiles, birds and mammals. The early amniotes were reptiles that differed from the amphibians in having better jaws, strong skeletons, and slender limbs, making them much more agile than the amphibians. These first amniotes probably looked like primitive lizards. From them diverged the lines that later produced turtles, crocodiles, lizards, dinosaurs, birds and mammals. Some later stages in the story of the vertebrates, during the Permian and Triassic, are best recorded in South Africa, in the Karoo escarpments of the eastern Cape Province.

A century and a half ago, Charles Darwin's African colleague, A.G. Bain, was building roads around the little military town of Fort Beaufort. From Fort Beaufort the roads north climb up an escarpment, cutting through sedimentary rocks, which are of Mesozoic and Permian age. The Katberg Pass, built by Bain, is one of Africa's finest: a route chosen with a geologist's eye. Almost intact in its Victorian form, it is a magnificent ascent through a precious remnant of cloud forest.

In these hills Bain discovered a treasure store, filled with fossils of monsters that he recognized as early reptiles. He eventually shipped them to the British Museum where they were examined by Sir Richard Owen. One of Bain's early discoveries was an enormous beast called a pareiasaur. This animal, which ranged up to 3 m (10 ft) long, had a massive skull, and teeth like modern

Figure 9.7 *Dimetrodon*, a big flesh-eating animal from the early–mid Permian of Texas. It had a huge sail on its back, and long sharp teeth. Length 3.5 m (12 ft) (from collection of the Royal Tyrrell Museum of Palaeontology/ Alberta Culture and Multiculturalism, Canada).

Figure 9.8 *Cynognathus*, a mammal-like reptile. This animal, called a cynodont, was a wolf-sized predator. Its skull and teeth were dog-like (Cynodonts were carnivores; Dicynodonts, a different line, were plant-eaters). Early mid Triassic (courtesy of Department of Geology, University of Saskatchewan).

Figure 9.9 *Morganucodon*, a small rat-sized early mammal that lived in the late Triassic (courtesy of Department of Geology, University of Saskatchewan).

herbivorous lizards: it probably ate plants. Another discovery was a beast he called a 'Bidental', after its two major tusks, later renamed *Dicynodont* by Owen. These were mammal-like reptiles, plant-eating animals ranging up to the size of a rhinoceros. One of the dicynodonts from South Africa was a beast called *Lystrosaurus*. This animal was a herbivore. Its fossils have also been found in Antarctica and China. Because *Lystrosaurus* was much too ponderous to have been capable of swimming from Africa to Antarctica, the only possible conclusion is that the continents were once joined, as du Toit had surmised. Appropriately, one of the major faults in the ocean floor south-east of Africa, which formed as the continents split apart, is called after Bain.

Other animals collected from these strata were carnivores that presumably ate the herbivores and also any eggs, amphibians or fish they could catch. Bain and his collaborator Owen correctly recognized that the African fossils were similar to mammals and quite unlike dinosaurs. Primitive mammal-like reptiles fill the collection – there is no sign of the dinosaurs. But wait, is there not a problem here? We all know that the dinosaurs came before the mammals. How then can these obscure African hills contain abundant mammal-like reptiles? Is not the story muddled up? Bain, a century and a half ago, recognized this problem: the European discoveries had produced powerful, terrible dinosaurs, and yet here was a community of animals much more closely related to ourselves.

The answer is that these rocks in the Cape are *older* than the dinosaur-bearing strata of Europe and America. The amphibians were succeeded in the dominance of the land by animals that laid amniotic eggs, but initially the dominion over the land passed *not* to the dinosaurs but to our own ancestors, the mammal-like reptiles. Dinosaurs came later. During the Permian and Triassic it

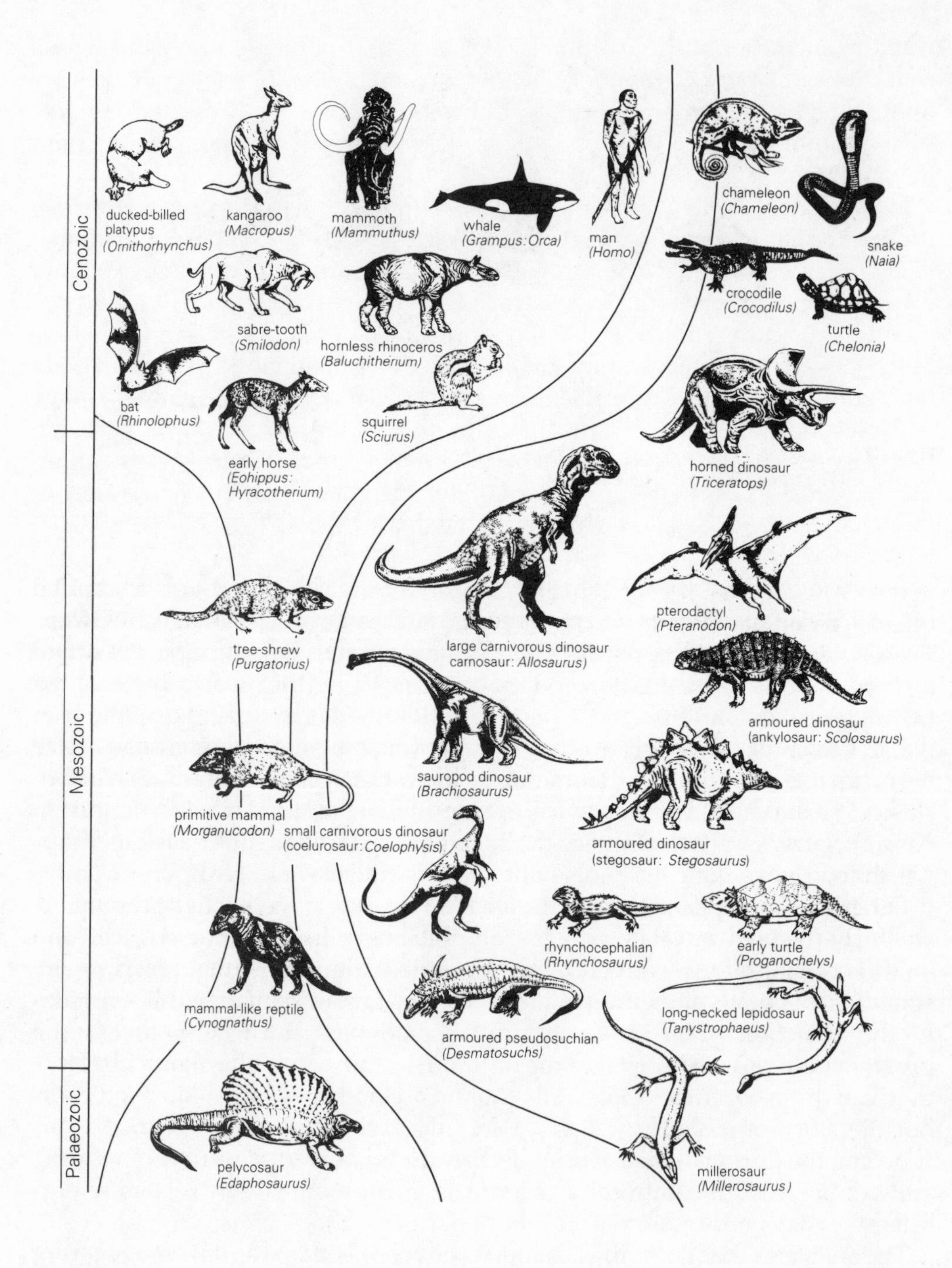

Figure 9.10 Life in the Mesozoic and Cenozoic (courtesy of Department of Geology, University of Saskatchewan).

was the ancestors of the mammals, our line, that populated the land. Fossil collections of Upper Carboniferous, Permian and early Triassic vertebrates are dominated by the remains of the mammal-like reptiles. They formed a widely varied community, obviously based on a very diverse vegetation. We are their descendants.

In the late Permian and the Triassic, these animals, called therapsids, progressed far toward mammalian character. The earlier species of mammal-like reptiles were egg-layers, although their descendants today bear live young. Bearing live young is not unique to mammals: the coelacanth has already been mentioned, and *Peripatus*, tsetse flies and some snakes do the same. There is much evidence that the later mammal-like reptiles maintained a constant body temperature, and they were able to breath steadily, even during eating, as they had a secondary palate in the roof of the mouth. Dinosaurs, like birds, probably had gizzards: in contrast, mammal herbivores need to chew more when eating. The mammal-like reptiles probably had whiskers, and perhaps hair. Many of them had feet with equal toe lengths, a first digit with two joints and four digits with three joints, as in our own hands.

The mammal-like reptiles constituted only one of several groups descended from the early reptiles that evolved the amniotic egg and so managed to escape the last tie to the water. One of the early distinct reptilian groups still extant includes the tortoises and their relatives who have returned to the sea, the turtles. Also still with us are the scaled reptiles: lizards, snakes and the like. Some of these early reptiles evolved in extraordinary ways: one even had a neck to match a giraffe. One of the more obscure reptiles is the tuatara, *Sphenodon*, which still survives, but is now restricted to a few scattered offshore islands in New Zealand. The first relatives of the tuatara appeared in the Triassic, perhaps 225 million years ago. They look like lizards, but have very different skulls: modern ones have relict third eyes, and teeth fused to the bones of the jaw. A distant early relative reached the size of a pig; many others appeared, lived and died. Today, of the whole collection, only the tuatara has a tenuous survival, recently driven off mainland New Zealand by humans and introduced predators, but hanging on in old age (they, incidentally, live as long as people) in the fragile security of a few islets.

FOOD

By the heyday of the mammal-like reptiles, during the Triassic, a complex community of life existed on land. The food chain on the land was not very different from the modern (or at least the pre-human) pattern in Africa or South America. At the base of the chain were the bacteria. They ruled the world in the Proterozoic, but in a sense they still dominate the planet today. Some maintain their independent existence, others depend upon more complex eukaryotes to carry them around; yet others have symbiotically become a part of those eukaryotes. But even today it is still a bacterial world – the basic pattern of gathering and using energy and resources has not changed since the late Archaean. From the bacterial point of view, if there were such a thing, a eukaryote's body plays the same role that a machine does for us. It may provide transport, or a home, or a powerful new way of collecting food. Consider the position of a bacterium in the stomach of an elephant. It has a nice warm home at a steady temperature; enormous quantities of food pass through daily; there is, eventually, the prospect that the bacterium's descendants can colonize the gut of the next elephant calf to come along. In short, perfection; how much more attractive than an uncertain future in a hydrothermal spring given to wild fluctuation, explosion and extinction.

Bacteria and their descendants manage the disposal of waste on the planet, whether in the mammalian gut, or symbiotically in plants, or in soil, or in mud. There is a world-wide oxygen-poor environment on the sea floor, immediately below the sediment surface, which is primarily bacterial territory. Assisting them, of course, are the worms, termites and other things that burrow, but many of these too have bacterial symbiotes. At Steep Rock Lake, in Ontario, Canada, it has been suggested that termites penetrated more than 300 m (1000 ft) below the surface: they had bacteria to help them, digesting the food in their guts.

Equally important is the management of nitrogen. Much of this is now done by bacteria associated with plants such as those attached to the roots of peas and beans. Nitrogen is essential to proteins and nucleic acids, and most of the nitrogen in our bodies has been extracted from the air and made chemically useful, or fixed, by bacteria; we get the nitrogen second hand (or, more probably, after hundreds of recyclings). Only within the industrial age has humanity learned how to fix nitrogen from the air, in fertilizer factories.

The first visible base of the food chain today is the plants, including the oceanic plankton. These, with their chloroplasts, are but cyanobacteria writ large. It is the plants that handle the carbon-dioxide reprocessing of the planet. Within the plant families there is extraordinary interdependence as well as competition; one plant will grow on another, or use the compost provided by the death of a third. In many tropical rainforests, the plants have been so successful that virtually all the soil nutrients have been taken out and placed into the living community; plants must gather nutrients from each other or the air and rain and not from their roots. The primary resource, sunlight, is well utilized: little light actually reaches the ground. Furthermore, to a great extent a large rainforest makes its own weather. Water is evaporated, forms clouds, falls again and continues the cycle. The actual input of new water from the ocean may be only a third or a quarter of the amount cycled. Dryland forest, too, has a marked impact on the colour and hence the reflectivity and temperature of the land: in recent years we have made enormous changes to the colour of the dry tropics by chopping down trees, and have probably altered the pattern of rainfall in these areas.

The management of rainfall by the biosphere probably began in the Carboniferous. The swampy coal forests included tall trees, perhaps with a high forest canopy, under which flourished a series of lower storeys of plants, inhabited by diverse insects. A modern parallel is in the swampier parts of the tropical rainforest: parts of Amazonia are regularly deeply inundated. Since the Carboniferous, the forest zone may have acted similarly to the modern rainforest, as a weather factory influencing rainfall distribution.

Today, most land animals are insects. On the modern Earth there is an enormously complex interrelationship between insects and plants, but this is mostly dependent on flowers. In the early forests, flowers had yet to evolve. Nevertheless, there was almost certainly a complex web of food processing, with small insects eating live and dead plants, and larger insects preying on them. Today, on Little Barrier Island, a protected island off the coast of New Zealand, the main role of scavenger is played by an extraordinary large bug, several centimetres long, called the weta; this beast is the insect equivalent of the mouse. During the Carboniferous, much of the scavenging was probably done by insects behaving like the modern weta. Higher up the Carboniferous food chain came the amphibians, which ate insects and each other.

By the mid-Triassic, things had completely changed. The mammal-like reptiles had evolved a much more complex, essentially modern, land food chain. In place of the pattern of the fishes – each fish a predator and each fish predated upon by other fishes – the land animals divided into a much more distinct class of grazers and a class of predators. The bulk of the mammal-like reptiles were probably herbivores. Only a small proportion of the population was carnivorous. The modern African fauna are similar, with small numbers of lions, cheetah and hunting dogs preying on large herds of herbivores.

Thus we have, by the Triassic, a modern-looking world, with a complex and diverse plant life, a variety of mammal-like herbivores and a small population of mammal-like carnivores. By about 200 million years ago some animals had so evolved that they looked like modern mammals: the transition had been made to animals that would not appear out of place today. Bain's bidental monsters and their relatives were becoming a modern community. All seemed set for monkeys and humans to appear.

But we lost the battle. Something surprising happened. The mammal-like reptiles did not inexorably evolve into nice warm furry creatures and then to apes and people. Instead, they almost disappeared. Better, more successful animals came along, and they were not mammals. The ruling reptiles took over: the dinosaurs. Our ancestors were reduced to a few small creatures scurrying around in the undergrowth.

FURTHER READING

van Andel, Tj.H. 1985. *New views on an old planet: continental drift and the history of the Earth.* Cambridge: Cambridge University Press.

Lambert, D. and the Diagram group. 1985. *The field guide to prehistoric life.* New York/Oxford: Facts on File Publications.

Lillegraven, J.A., Z. Kielan-Jaworowska and W.A. Clemens (eds) 1979. *Mesozoic mammals – the first two thirds of mammalian history.* Berkeley: University of California Press.

10
The rule of the reptiles

THE REIGN OF THE ARCHOSAURS

In the Triassic, change came: the archosaurs or ruling reptiles appeared, and many of the mammal-like beasts became extinct. Perhaps today's most familiar large carnivorous reptile is the crocodile. One of the diversions of geological field work in Africa is to sit hidden on a river bank, very still, and watch for what are locally called 'flat dogs' or, in deference to Western consumer tastes, 'handbags'. The signs are obvious – little mounds of fish teeth, 'croc drops' – but the animal is not obvious. At first, all that can be seen is the spouting and humping of hippos, but eventually two nostrils appear, if the scene is peaceful,

Figure 10.1 The Mesozoic landscape, from the pterosaur's point of view. Some pterosaurs were huge, ranging up to 12 m (40 ft) wingspan (painting by Vladimir Krb, Royal Tyrrell Museum of Palaeontology/Alberta Culture and Multicultualism, Canada).

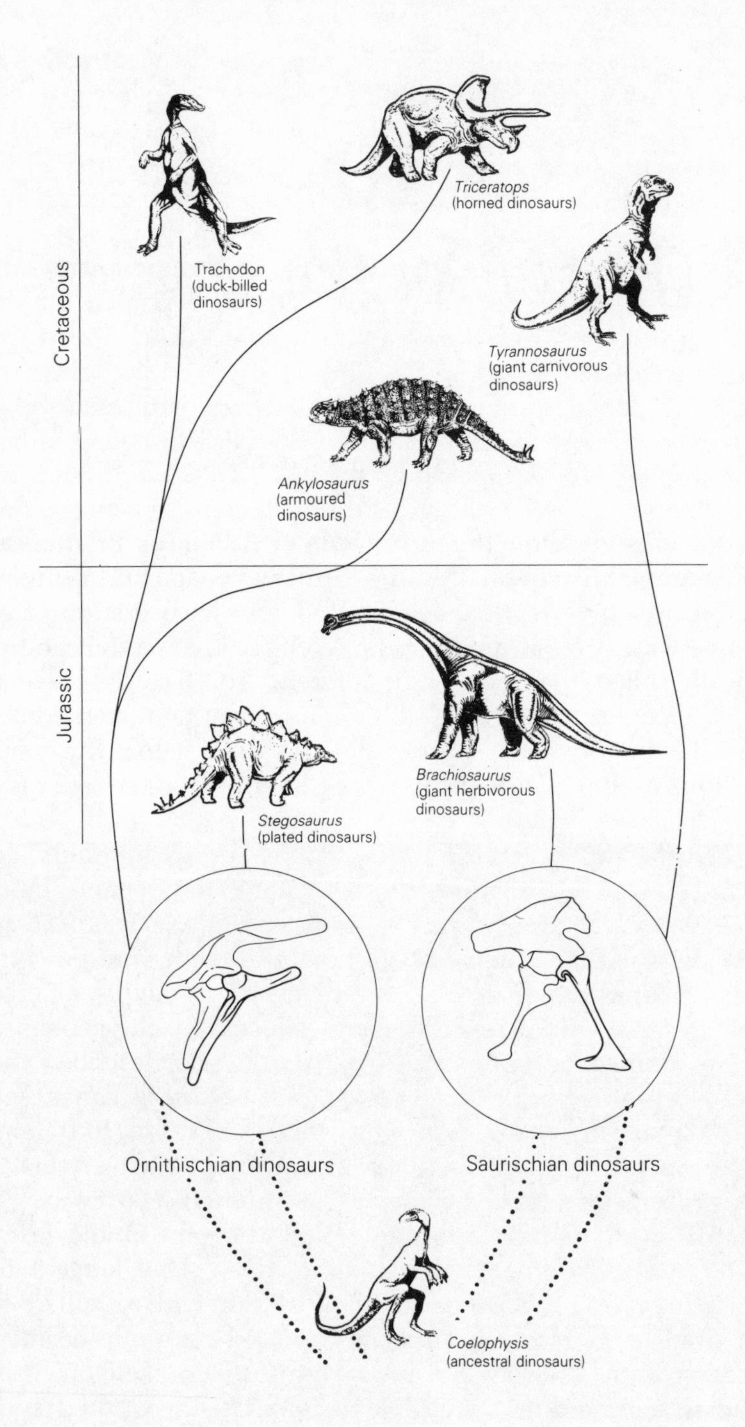

Figure 10.2 The two types of dinosaurs: bird-hipped (Ornithischian) and lizard-hipped (Saurischian) types may have evolved from a distant parent reptile, although the closeness of their relationship remains controversial. Birds, ironically, are likely to have come from the Saurischian line (courtesy of Department of Geology, University of Saskatchewan).

and then slowly the eyes and head of a wary reptile float across the pool. Crocodiles are cunning and superb at concealment; they are also marvellous parents, carefully taking up their young in their mouths to move them (from which action comes an old, false, accusation that they are cannibal). Somehow the crocodiles have survived: other reptiles went on to greater things, and perished.

The best known of the ruling reptiles of the Triassic are the dinosaurs. Dinosaur fossils were discovered in the 1820s, and named by Sir Richard Owen. The name comes from the Greek for terrible (dino-) lizard (-saur). The oldest dinosaur fossils are from the end of the Middle Triassic in South America. Unfortunately, the late Triassic fossil record of dinosaurs is very poor. There were two main orders of dinosaurs – the bird-hipped Ornithischia and the reptile-hipped Saurischia. The ornithischians were herbivores; the saurischians included carnivores and herbivores. There is much controversy about the exact relationship between the two groups of dinosaurs. The saurischians can be divided into two-footed carnivores, and four-footed herbivores, but it is not clear how closely related these two groupings were. The ornithischians share a common ancestor, which may have been a primitive saurischian.

THE SAURISCHIA

Saurischia became common early in the history of the dinosaurs, and dominated the land ever after, until they suddenly died. They began as bipedal (two-footed) flesh-eaters. In the Jurassic, a typical example was *Allosaurus*, 9 m (thirty feet) long, with a massive skull, powerful teeth, and powerful claws. One of the first dinosaurs to be discovered was *Megalosaurus*, a fierce Jurassic flesh-eating creative whose remains were collected by the geologist William Buckland and named by James Parkinson (who also described the disease named after him). A Lower Cretaceous example of a saurischian is *Deinonychus*, found in Montana. This animal was lightly built, about 3 m (10 ft) long, weighing about as much as a large man. It could probably run very fast, and was equipped with ferocious curved claws, to disembowel its prey.

Perhaps the most fearful dinosaur of all was one of the last, the late Cretaceous *Tyrannosaurus rex*, whose name means 'the king of the tyrant lizards'. *Tyrannosaurus rex* was nearly 6 m (20 ft) tall, with a skull up to a metre and a half (5 ft) long, massive hind legs and a beautifully balanced tail. It weighed seven tons. Viewed in a properly mounted skeleton, it is a finely designed beast, agile and taut, well able to dominate the world. It is often seen as a symbol of a terrible age, though we forget that both physically and in intelligence the sperm whale is a more powerful predator than *Tyrannosaurus rex*, and the sperm whale is itself hunted.

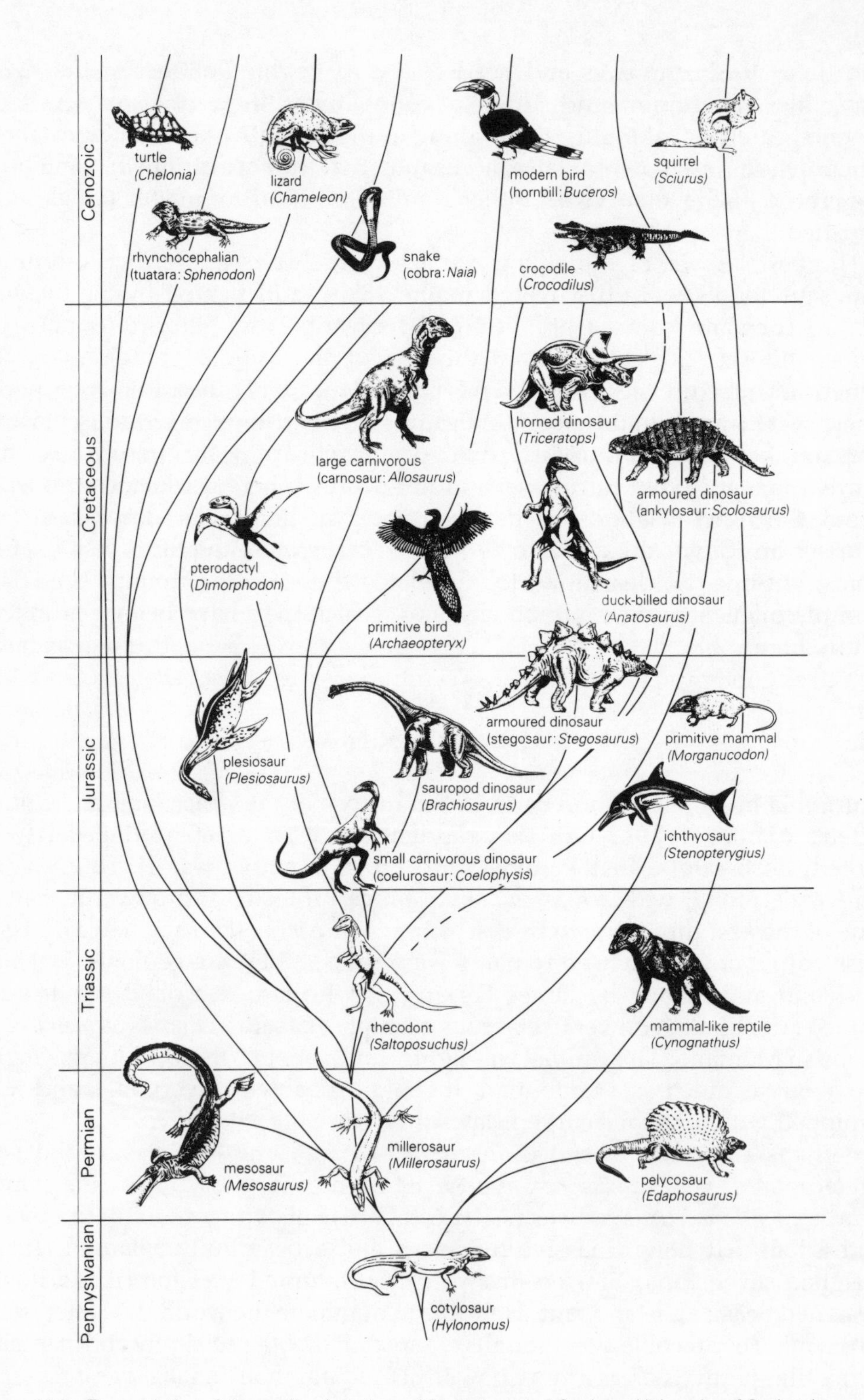

Figure 10.3 The evolution of the reptiles (courtesy of Department of Geology, University of Saskatchewan).

Another group of saurischians produced plant-eating herbivores – *Diplodocus*, *Apatosaurus* (*Brontosaurus*) and *Brachiosaurus*. Some of these may have been like modern giraffes, strolling along cropping the tops of trees, while their less lofty relatives browsed the lower strata of vegetation. These herbivores were mostly large, 10 m (30 ft) long, or more, but some carnivorous dinosaurs were small, even rabbit size.

THE ORNITHISCHIA

The ornithischians, in contrast, were all herbivores. The first to be found as a fossil was the celebrated *Iguanodon*, supposedly discovered in some road stones by Mary Ann Mantell during a walk on a beautiful English spring day. The discovery caused a sensation, as the savants (especially Sir Richard Owen) tried to reconstruct the animal. For human society, the dinosaurs had arrived.

Many ornithischians developed armour. An early example is *Stegosaurus*, from the Upper Jurassic of North America. *Stegosaurus* had a small head and little brain, but was protected by vicious spikes at the end of its long swinging tail and a double row of armour plates along its backbone that would have served as good protection, as well as being well suited to radiating excess heat. The most powerfully armoured dinosaurs, however, were the ankylosaurs, which may have evolved in Europe and then spread to North America and Asia, where they were common in the Late Cretaceous. Ankylosaurs had heavy limbs and were four-footed, to support a trunk that was entirely covered

Figure 10.4 *Stegosaurus*, an Upper Jurassic herbivore. This animal had two rows of plates along the neck and back, and spikes on its tail, which would have been a powerful defensive weapon, swinging into any attacker. In most reconstructions the plates are shown upright; they may also have sometimes lain flat as shown here. Length up to 9 m (30 ft) (courtesy of Department of Geology, University of Saskatchewan).

by small interlocking bony plates. The animals looked rather like giant armadillos, a tough lunch for a tyrannosaur.

Another large group of ornithischian dinosaurs was the hadrosaurs, or 'duck-billed' dinosaurs. These were bipedal herbivores, and were especially common in the Upper Cretaceous of North America. They may have migrated across the land in large herds, moving to higher land to nest. The North American Upper Cretaceous also contains abundant fossils of horned dinosaurs, or ceratopsians, such as *Triceratops*. These were quadrupeds,

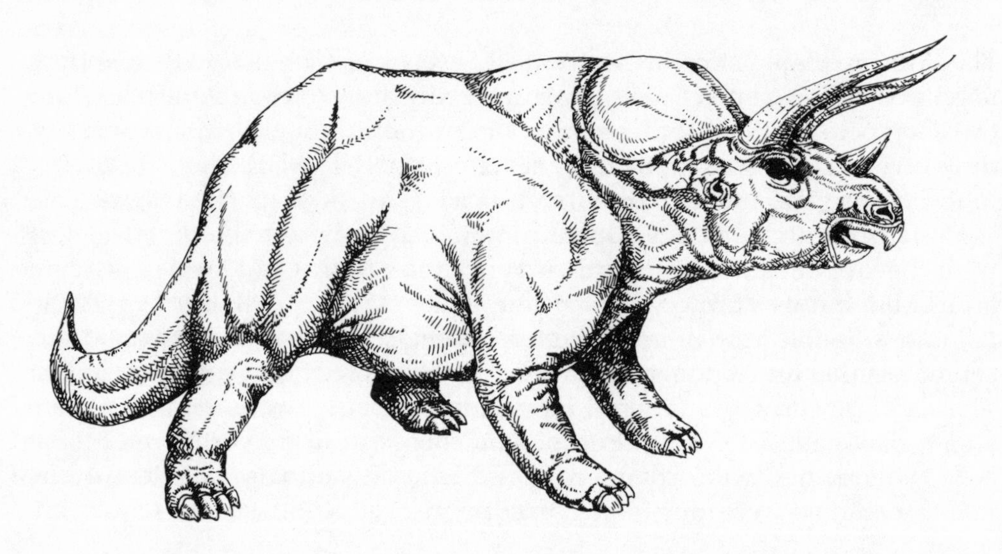

Figure 10.5 *Triceratops*, a late Cretaceous herbivore, about 7 m (24 ft) long (courtesy of Department of Geology, University of Saskatchewan).

browsing on bushes and trees, and were the equivalent in the Cretaceous of a modern black rhino. Their skulls were heavily armoured, with horns to protect the nose and eyes. *Triceratops* may have grazed in herds. When a predator, such as a tyrannosaur, appeared the *Triceratops* herds could have formed defensive circles, just as musk-oxen or buffalo do today, with the armoured heads outwards: a difficult and dangerous prey, even for a tyrannosaur.

THE DOMESTIC ECONOMY OF THE DINOSAURS

Dinosaurs derive their name from the Greek words for terrible lizard. The name evokes an image of a cold-blooded reptile, sluggish at night and in cold weather, active in the warmth of the day, and with a ferociously selfish lifestyle that admitted no domestic society. In contrast, we think of mammals as warm, social creatures.

There is strong controversy about these images. The closest surviving relatives of the dinosaurs, the birds, are warm, fluffy and often social. It is probable that dinosaurs were warm-blooded. One of the most striking characteristics of the dinosaurs is their body size – although small dinosaurs did exist, many were huge. Large animals have a low ratio of surface area to volume, and so lose bodily heat much more slowly than do small animals. An animal with a diameter of a metre (3 ft), living in a subtropical climate, would maintain a nearly-steady body temperature without any special effort, simply because of its size. A large dinosaur would therefore have a steady temperature, even with a reptilian metabolic rate. In contrast, some small mammals have to spend up to 90% of their energy simply in maintaining a constant temperature: being a mammal is very expensive. The disadvantage of using large size to maintain a steady temperature is that the animal is very vulnerable to extremes of temperature that persist for long periods; a cold snap of over a week could be fatal.

Baby dinosaurs, being smaller, would have been less able to maintain a steady temperature. They probably grew rapidly to offset this problem, and may have eaten a diet rather different from adults of the same species. In some species, the adults probably never stopped growing, like modern giant tortoises. Some adult herbivorous dinosaurs seem to have had a social structure, nesting together and tending their young. The weight of the eggs ranged up to 5–7 kg (15–20 lbs). Some may have migrated extensively, following the seasons and the availability of food, just as elephants used to migrate in Africa. By the end of the Mesozoic, many flying animals had evolved, and some of these may also have migrated, exploiting their ease and speed of movement.

FLYING REPTILES – PTEROSAURS AND BIRDS

That pigs may fly is excellent, it has been said; that pigs might fly is intriguing; but that pigs actually will fly is non-newtonian and may make a nasty mess on the pavement. Yet should we and all other animals abandon this Earth to pigs we would, some millions of years hence, see a planet inhabited by flying pigs (as well as pigs in trees, swimming pigs, herds of grazing pigs predated upon by pig-lions and, perhaps, an animal farm of pigs in politics). Evolution, through natural selection, is capable of producing diversity to fill all the available ecological niches: through evolution, many unlikely pigs have taken wings.

Flight has evolved many times. Fish do it, and are a marvellous sight in the tropical ocean at sunset with dolphins chasing behind, leaving a phosphorescent wake. All manner of beetles do it, ants do it, bats do it, people do it. The first land vertebrates to fly were the pterosaurs, which appeared in the late Triassic. They were not dinosaurs. The early pterosaurs were probably fine flyers. Pterosaurs became widespread and varied in the Jurassic, including the

Figure 10.6 *Dimorphodon*, an early Jurassic pterosaur. Wingspan about 1.5 m (5 ft) (courtesy of Department of Geology, University of Saskatchewan).

pterodactyls that, after the horrors of Sir Arthur Conan Doyle's tale '*The lost world*', have so exercised the imagination of Hollywood. Towards the end of the Cretaceous, the pterosaurs became less diverse, with the extinction of many species, but some of those that remained became very large, including *Pteranodon*, with a 7 m (25 ft) wingspan, and *Quetzalcoatlus*, which had a wingspan of 11–12 m (up to 40 ft) and was as heavy as a man. Some of these larger species may have behaved much as albatrosses do today, ranging across the seas. Many were fine flyers, with a rudder forwards on their heads, rather than a tail at the back, as in modern birds. The larger species would have had problems getting airborne, and may have relied on large waves or cliff-edges. They may have been warm-blooded and covered with sleek hair, not Hollywood monsters but warm furry things, possibly white to maintain their heat balance.

Pterosaurs flew for many millions of years before the dinosaurs discovered the advantages of flight. They produced the birds, which seem to have been better adapted to flight than the pterosaurs, because they displaced them. The ancestral stock of the birds is debatable – suggestions even include the crocodile line (no less extraordinary than pigs), but it is most likely that birds came from a dinosaur or a relative of the dinosaurs. Indeed, it can be argued that they *are* dinosaurs: we really harbour dinosaurs in our budgie cages.

The probable link between dinosaurs and birds is in the most precious of all fossils, *Archaeopteryx*, which is the oldest clearly feathered fossil bird known. It lived in the late Jurassic, about 150 million years ago. Of all the missing links, this is the best. It is a mixture of reptile and bird, with feathers and wishbone, and also a long bony tail, teeth and three fingers on its front claws. It is a dinosaur, but it lies on the evolutionary path between its archosaur ancestors and modern birds, which have, over their evolution, (except the young of the modern South American hoatzin bird) lost their teeth and front claws.

Figure 10.7 *Archaeopteryx*, the first known bird. Late Jurassic, about 1 m (3 ft) long (courtesy of Department of Geology, University of Saskatchewan).

Birds keep their legs free for walking or running, or – one of nature's most dramatic sights – for an eagle's fall, killing prey. They seem to have had advantages over the pterosaurs, and so the second reptiles to fly displaced the first. From them we have our modern birds: the dinosaurs are still with us, as chicks; warm, cuddly, fluffy little things, executed horribly, then roasted or fried by us.

PAINTING THE LAND: THE ARRIVAL OF COLOUR

We may eat dinosaurs, but what did the dinosaurs eat? During the Cretaceous a change came over the face of the land. The Jurassic vegetation had been a mix of cycads and horsetails, ferns, conifers and ginkgos. Fragments and small communities of this cycad-rich vegetation still exist today, in obscure places such as little stream valleys in Africa. Many of the plants were **gymnosperms**, a group of seed plants that do not have flowers. Today, their larger-scale descendants are present in the tall, dense forests of southern Africa and New Zealand, which are filled with trees known as podocarps and have abundant tree ferns. The conifers too have descended from this time, and they still dominate the north of the planet, competing vigorously with deciduous trees and making up the bulk of the Canadian and Siberian forests.

But from our point of view these are boring trees. Something is missing. We would call it 'aesthetic appeal'. A gymnosperm forest is a place haunted by the ancient subtle beauty of dark greens and russets. One of the finest of all trees is the African yellow wood, an olive-leaved and elegant, slow-growing cloud forest podocarp that can reach over 30 m (100 ft). But it is monochrome, and it is beautiful only when set against a lush background of colourful cloud forest species. By itself, it is boring, and there is an evolutionary reason why we find it so drab.

During the early Cretaceous, flowers evolved. We delight in the flower because it is aesthetically appealing to us, and that is precisely why it evolved. Our distant ancestors were drawn to it because it implied a site to be remembered – nectar was there, fruit would follow. Flowers are subtle. Humans like them because our senses tell us that they smell sweet and look pretty. They have entered into a symbiosis with us. For example, the extraordinary flora of the Cape of Good Hope have spread across all the inhabited continents (often as pests) simply because they managed to attract Dutch and English gardeners who made them the dominant part of the 'traditional' flower garden around European and American houses.

Flowers were evolved to take advantage of the abilities of animals, especially insects. Plants could use insects to carry pollen for them. Wind and accidents of gravity are not especially reliable means of genetic dispersal. Some plants use other strategies – pine cones can explode in fires – but the best way of dispersing one's genes is for one or both sexes to travel, often the male, because the female is more likely be heavy and to need the security of staying in one place. Animals have the same problems in genetic dispersion. Many ants make the male fly away, many mammals drive young bulls out of the herd, and humans have evolved an elaborate collection of rituals, varying from the arranging of marriages between different villages, to its modern substitute, the university.

Plants are rather more restricted in this business of the birds and the bees, not being as mobile as most university students, so they use insects, birds, bats and people to do their work for them. Hence the flower, which attracts the birds and the bees: things with wings then carry the pollen to the flower of a distant plant, which receives the distant genes. As the process becomes more sophisticated, bird and bee species become specific about which plant type they visit, for it is advantageous to them if the plant prospers. Thus hummingbirds favour scarlet runner beans (they like red), and bees are very particular about plant and season, whether they go to clover in Saskatchewan, rewarewa in New Zealand, leatherwood in Tasmania (in all of these places the honey bees are introduced) or introduced jacaranda in Zimbabwe (where the bee is native but the tree is not).

A flower is a wonderful device. The fig is a good example. Some species of fig have as many as twenty dependent insect species, many of which actually live inside the fig. Among these are parasites, but there is a very fine line between parasite and inhabitant, and between inhabitant and symbiont. As it is in the interests of the parasite to promote the interests of the plant, the parasite soon evolves a symbiotic lifestyle, helping to fertilize the seed. In consequence, figs have an intricate flowering process that depends on live-in insects and produces various different stages of food for them. The result is that some species of fig tree bear fruit all through the year. Reams could be written about Christ's comment to the fig tree before Jerusalem: it has a complex Talmudic intricacy about it which surely would have delighted Rabbi Gamaliel when he examined his breakfast figs for bugs, which he probably did. Flowers mean symbiosis, co-operation.

Once flowers had evolved, they spread rapidly and plants began to compete by colour and scent. The flowering plants, the **angiosperms**, came to dominate the central parts of the planet, and the Earth took on colour and smell. The insects were the partners in the change: mosquitoes, wasps, bees, butterflies, moths and assorted other bugs – the spread of the flowers was matched by the variety of insects serving them. Birds, too, exploited the flowering plants, and dinosaurs such as hadrosaurs ate them. All this, of course, implies that insects and animals were able to see and to smell the flowers. Vision and smell capable of distinguishing colours and scent probably developed at some time in the Cretaceous, when competition began between the flowers.

We have here an example of competition generating co-operation. Flowers are a competitive device, by which one plant can gain fitness compared to another. In response, insects compete to exploit the opportunity. Yet what is produced is a framework that offers new prospects to all. The plants prosper, the insects flourish, and to our own eyes, because we use senses and concepts which themselves have an origin in this process, the whole scene is filled with beauty. Here is the paradox of natural selection. In the immediate sense it is a dirty, deadly business – eat or be eaten, kill or be killed, exploit or be exploited, struggle ruthlessly to survive, or become extinct. Yet in nature the Darwinian competition is linked by an integral co-operation, at levels ranging from the fertilization of the fig to, perhaps, the management of the atmosphere. This collaboration is at the heart of life itself: we would die without it, all of us, from bacterium to astronaut. Collaboration is the other side of competition: the insect fertilizing a plant, or the dog-baboons that will deliberately attack and kill a leopard, and themselves be killed, to protect their family group, or the bird picking a crocodile's teeth are all collaborating, even if it is for good competitive reasons.

THE DINOSAURS, SCIENCE AND DINNER

The discovery of the dinosaurs fuelled a ferment among the early Victorian palaeontologists, such as Buckland, Mantell, Lyell, Cuvier, and especially Owen. Their work helped to make experimental scientific research a permanent part of the structure of Western society. The result of this palaeontological ferment among the geological savants (the word scientist was not yet common), after Charles Darwin's work, was a change in the world-view of society as a whole. The scientists had looked at the Earth, in the fossil record and in living communities, and unexpectedly had been taught much about the nature of humanity. But the new knowledge brought evil too: the doctrines of social darwinism and eugenics, that try to make human society conform to what was seen as the natural mould, led to racism and social wrong. These ideas helped to lay the foundations for Nazi philosophy.

Some of the stories from the early, more innocent, days of research illustrate the real scientific method. Successful science is not carried out according to a rigorous regime set down by philosophers. Instead, it is a combination of experiment, muddle, mistakes and imagination. For instance, the remains of *Iguanodon*, described by Mantell, provoked an attempt at reconstruction. The result was a large and imaginative model of a quadruped. It was very incorrect, but it was the first step towards imagining Cretaceous reality. Several such models were built, of different extinct beasts; the leading savants of the day celebrated by having dinner in the hollow body of one. These extraordinary reconstructions can still be seen at Crystal Palace, in south London.

Dinner, in whatever form, is something for which dinosaurs and palaeontologists hold a common respect. One of the most important early geologists, Dr Buckland of Oxford, was famous for his claim to have eaten all possible foods – when shown the heart of the King of France (which had come, somehow, to England), he gulped that down too, perhaps in the spirit of research. People, unlike Iguanodons, are omnivores, but the diet is not always right for them.

Bain, discoverer of the bidental mammal-like reptiles (which he sent to Owen), was equally concerned with the matter of dinner. On receiving some sausages sent by a colleague, Dr Borcherds, he replied in doggerel, illustrating the processes of geological deductions. Perhaps the last stanza also alludes to Mrs Mantell? He gave thanks:

'. . . for your kindness and pains,
In sending such precious *organic remains*;
In vain for description of them you may try all
The pages of Buckland, of Mantell, of Lyell;
For like our bidentals they must be unique,
Only known to our geological clique.
In science a novelty better by far
Than glyptodon, mammoth or famed ichthyosaur;

But first it behoves me to fix their position,
Whether *Pliocene*, *Miocene*, or of transition,
For which I'm as able, au fait, and as knowin'
As e'er Cuvier was, or yet Dr Owen!

They cannot be *Secondary*, for they're first-rate;
They're prime, though they do not as *Primary* date;
Fat and flashy-like cubes, with saliferous particles,
And dark *peperinos* compose all these articles.

No *ox-hides* occur, but a thin dermal casing,
The varied contents of the fossils embracing;
So among the *Conglomerates* I must enlist'em
For I'm sure they belong to the *Bologna* system!

Other fossils emit a strong smell it is true;
These not only smell sweet, but have a rich *gout*.
They're thoroughly *gneiss*, yet no feldspar or quartz,
Or mica compose their constituent parts.
They have but one *fault*, which I'm sorry to say
That with me they'll be subject to sudden decay.'

Kind regards to your lady, without whose kind aid,
These relics had ne'er to the world been displayed.
Farewell, my dear Borcherds, in haste I remain
As truly as ever, your's A. Geddes Bain.

27 May 1844.

The palaeontologists' concern about food is central, because finding food – gathering energy – is one of the chief businesses of any living form (along with the avoidance of becoming food, and the urge to reproduce). No doubt the Victorian savants were not the first to dine in the belly of an Iguanodon (an excellent repast for a megalosaur), but they were probably the first to choose so to do. As that generally excellent guide, 'The Restaurant at the End of the Universe' (part of the *Hitchhikers guide to the galaxy*) points out, the question of 'where shall we have lunch?' is something only the privileged few of a unique species can consider. Dinosaurs had no such choice.

THE LATE CRETACEOUS DISASTER

We have seen how, sometime in the Triassic, the dinosaurs took over as the dominant large animals on the land, displacing the mammal-like reptiles. Before that, the ancestors of the dinosaur had been a minor part of life on land. They were carnivores, playing a role in the land ecology but not having a major part in the economy of the world. Afterwards, dinosaurs conquered the continents, and the mammals were restricted to a minor role, scavenging for insects and the like, rather as opossums today scuttle around the dark places and basements of North America. A small mammal-like jawbone was found and studied by Buckland and Cuvier, from the deposits that had produced the discovery of dinosaurs.

By the late Cretaceous the dinosaurs had dominated the land for well over a hundred million years. A complex ecological web had evolved, inhabited by highly efficient social groups of herbivores and powerful predators, watched over by the birds, depending on a mixed vegetation including the flowering plants. One can visualize herds of duckbilled dinosaurs, ceratopsians and others, grazing on the bushes and forests, occasionally falling to *Tyrannosaurus* and the other great predators. Even today in Africa virtually no animal dies peacefully – death comes as a predator – but there seem to have been uncommonly powerful predators in the world of dinosaurs; it was a vicious place.

In the Upper Cretaceous, an interesting beast appeared, *Stenonychosaurus*. It has been found in the Judith River Formation in Alberta (not far from one of the finest dinosaur museums of the world, the Royal Tyrrell Museum of Palaeontology in Drumheller). *Stenonychosaurus* is fascinating because it had a brain that was large in contrast to most other dinosaurs, and which compares favourably in size with the brains of early birds and mammals. It has enormous eye bones (did it hunt at night?) and may be distantly related to the early birds, as it has many bird-like features in its skull. This beast, like all organisms, was well adapted to its contemporary environment. Dinosaurs were not necessarily stupid, not necessarily sluggish, not necessarily uncompetitive. By the end of the Cretaceous, somewhat later than the time of *Stenonychosaurus*, the dinosaur community was the most successful, most competitive, most ferocious collection of animals that had ever lived on Earth.

Some scientists have amused themselves by speculating on what would have happened had the dinosaurs been granted a fragment more time on Earth: say a million or two more years. *Stenonychosaurus* indicates that the late dinosaurs were evolving in the direction of becoming large-brained animals, with binocular vision and hands able to grasp and turn. Perhaps the dinosaurs were on their way to producing an animal with an intelligence comparable to humans. Given more time, they might have even produced a civilization. But something happened.

We do not know for certain what caused this great change, but it eliminated virtually every living thing on land heavier than about 25 kg (50 lbs) together with many smaller organisms. Around the globe, a layer of sediment rich in soot, unusually high in iridium, and with mineral crystals that must have formed by shock impact or high pressure, lies at the boundary that ends the Cretaceous. The soot seems to have come from huge fires. Every dinosaur became extinct, except for the birds. Across the land, especially in what is now North America (the best studied area), enormous changes took place in the vegetation. At sea, many of the tiny floating organisms (which become microfossils) were wiped out. The extinction was massive.

How sudden was the change? We do not know – the geological record on land is not very precise, and at sea it is subject to disruption by burrowing by worms, snails and clams and so on, so that the resolution, (or fineness) of most of the record is somewhere between 10 000–50 000 years at best, and usually worse than that. Any event that happens in a shorter period than this cannot usually be pinned down in the marine record because bottom sediment is reworked and mixed up on a small scale by marine organisms such as worms and molluscs. Perhaps the extinction took as long as, say, several million years, but there is a strong body of scientific opinion that it took 50 000 years or less. In geological terms, this is sudden. Are there any parallels in the geological record?

Interestingly, there are. The lesser parallels are in the frequent minor episodes of extinction that form a constant background noise on the geological record. Larger events occur too, some as great or greater than the event at the end of the Cretaceous. Massive marine extinction took place at the end of the Permian, and a phase of extinction also occurred when the mammal-like reptiles gave way to the ruling reptiles and the dinosaurs. We see the same process in other ways too. The wanderings of the continents have on occasion caused major changes in the pattern of life. Each time two isolated land masses join, the consequence is catastrophe for some species, which are wiped out by more competitive organisms from the other land mass. But the catastrophe that overcame the dinosaurs seems to have been immense, and probably also sudden. It was so widespread that it involved land and sea, eliminating animals, vegetation (a change in plants would be expected as animals change) and also much of the complex chain of life in the oceans.

We can imagine a modern analogy. Imagine the geological record of the present day and the next few hundred thousand years, as studied by a palaeontologist sixty million years hence. Perhaps our palaeontologist of the future, assuming humanity disappears, would be a marsupial, possibly descended from some New Zealand immigrant possum. Our future palaeontologist will see in our history of the present day something extraordinary, closely comparable to the terminal Cretaceous 'event'. About 100 000 years before the present day, something started happening in the Africa–Europe–Asia supercontinent. At that stage, if the observer is very lucky, the geological record might include one or two primate bones from Africa or South America: humans would be recorded as an obscure family of apes, possibly arboreal or nocturnal. Maybe a baboon would be preserved in the sediments of the Niger inland delta (its sea delta, along with the Nile's, having been deformed and metamorphosed in a continental collision long before the possums evolved a palaeontologist), but humanoid remains would be very rare.

Then would come signs of an ice age, massive vegetation changes and a wave of extinction linked with a few human bones. If the record were superbly detailed, the future palaeontologist would discover that mammoths and large animals went first, along with lions and elephants from North Africa, the Middle East and Europe (assuming that Mesopotamia, bits of the Sahara and the odd fragment of Greece's valley fill are preserved: most unlikely). The Americas might record the story of a similar swathe of extinction as early man suddenly swept from the Bering Strait down to the shores of Tierra del Fuego. Roughly three quarters of the genera of large mammals over 45 kg (100 lbs) in body weight became extinct in North and South America, probably through aboriginal hunting, though climate change may also have been important.

All this, of course, is most unlikely to be preserved in the geological record. More probably, the future palaeontologist would simply be able to say that many species of large animal were present a few hundred thousand years BC, and a few hundred thousand years AD they had all gone.

Perhaps the future palaeontologist would see more signs in pollen from the plants, preserved in sediment. About 3000 years ago the forests of the Mediterranean began to disappear. Then, as seen from the future, around the time of the 'final catastrophe' all the forest went. By 2000 AD, apart from a few islands of green and the remains of Amazonia, the vast pre-human forests had already mostly gone. Grasses replaced forest. Global climatic change probably accompanied the devastation, and innumerable species vanished – not just the obvious ones like the mastodons but the minute bugs and obscure plants which constitute the bulk of the fossil record.

To return to the viewpoint of the present day, it is clear that the ecological fabric has torn: even if humanity were to disappear tomorrow, the nature that would come back would be different from what was before. The change will be best recorded in sediments that are allowed to lie undisturbed, and the record of the devastation will mostly be in the shape of pollen grains and the like. To the future palaeontologist the pollen will record an inexplicable sudden change as forest species become extinct and grasses take over.

Another subtle record will also show in the marine sediments. There will be signs of radioactivity from material such as plutonium which we have made, but this will be minor and will fade. Assorted man-made chemicals will be longer lasting, but these too will eventually be destroyed. A brief explosion of marine organisms will be recorded in the North Sea, reflecting the enormous increase in the supply of nutrients, brought down by rivers from agricultural fertilizers and manure. But perhaps the most obvious record will be of the industrial metals that have reached the sea bed, especially the rare and chemically relatively inert elements (such as platinum and iridium). We have dug these up, used them and now scatter them again.

Will there be any other sign of our civilization? Probably not. The large cities will be washed to the sea. The sea floor will be recycled in subduction zones. The only likelihood of preservation of a man-made artifact (a beer can perhaps, or a Coke bottle?) is in an intracontinental sedimentary basin such as the North Sea (but that will probably be folded and metamorphosed when the Atlantic closes again) or perhaps Lake Eyre, in Australia. Our last relict may be a Foster's can. Just possibly the floor of Lake Victoria or Lake Chad will preserve the bones of an odd primate which seemed to have a large brain, but little sense.

All this is the doing of a few thousand years, well within the 'fuzz' of the geological record. And it is all the result of the actions of one species, from a previously minor line, that exploded across the planet. The future palaeontologist will have little chance of unscrambling the full story. Yet science is often lucky. Maybe our possum-palaeontologist, digging into his roots in some southern New Zealand lake sediment, will come across the bones of British weasels, or of a wapiti – or even a moose. That surely will give him a clue. Or will the question of how a moose swam the Pacific become one of the great riddles of palaeontology? The whole geological record is fragmentary and very difficult to decipher. That the dinosaurs blew themselves up by nuclear weapons is wildly improbable (though fun to write fiction about), but a major change of some sort did take place at the close of the Mesozoic era.

One explanation of this change is that a rapid but not catastrophic evolution took place at the end of the Cretaceous – say over a few million years – when some successful adaptation favoured the mammals and destroyed the dinosaurs. Perhaps the dinosaurs died out piecemeal, in the millions of years before the end of the Cretaceous, with the last survivors becoming extinct at the end of the period.

But the extinction affected the whole ecosystem, especially in the oceans. It was not confined to the dinosaurs. Possibly some organism developed a way of destroying a king-pin of the ecosystem, and all else crashed down in ruins. There is much debate about whether a complex ecology is more stable than a simple one – it probably is – but all sorts of apparently trivial events can undermine a complicated system. For instance, the Etosha Park in Namibia is one of Africa's richest game communities, but the whole ecological chain was devastated by a fence and the digging of a simple gravel pit, which uncovered anthrax that then infected certain key species. If a king-pin of the chain goes, all goes, and the vegetation of the planet also changes (as does the climate) until a new stability is found.

The conservative view is that the end of the dinosaurs was a change like many others in the geological record, more marked than most, but not exceptional (there were some earlier extinctions possibly of equal severity, especially in the sea). A small, but distinguished, minority of palaeontologists believe that at the end of the Cretaceous there was no marked extinction event at all.

The catastrophist view of the end of the Cretaceous is simpler. Something went 'bang'. An extraterrestrial impact is the favourite explanation – a large meteorite or comet could do the job, producing atmospheric disruption, fires, and a few years of sharp global cooling: enough to eliminate virtually all large animals and much marine life and to cause global fires, recorded in a world-wide soot layer. This impact hypothesis is supported by the unusual amount of iridium present in the sediments at the boundary between the Cretaceous and the Tertiary although several iridium-rich layers are now known, implying several events. We could probably do the same with a few thousand megatons of nuclear bombs and our industrial waste, leaving the puzzle for the possum palaeontologist.

Volcanoes also produce big bangs. Mt St Helens produced a little burp, but the eruption of Krakatoa was somewhat larger, a century ago. The explosion of Tambora in Indonesia in the early nineteenth century may have caused 'the year without a summer' in North America, and eruptions in Iceland have had an effect on the global climate. Santorini, a Greek island volcano north of Crete, destroyed much of the Minoan civilization and gave birth to the legend of Atlantis. This explosion occurred possibly around 1450 BC, although the date is disputed. Some geologists have suggested that its impact on humanity was not restricted to the Minoans: it may also have provided the cloudy pillar, helped set off the plagues and in its final explosion created the tsunami wave of the biblical Exodus. The hypothesis is controversial, not proven, but intriguing. More dramatic yet was the Taupo explosion of around AD 130 or 180 which devastated most of North Island, New Zealand. The debris of the eruption wiped out much of the life in the centre of the island, burying it in a layer of hot ash and lava particles. But all these are small compared to the once-in-a-million years event, and the once-in-a-billion years event is greater still. A really large

Figure 10.8 The tuatara, *Sphenodon*, (about 70 cm–1 m (2–3 ft) long), still living in New Zealand (courtesy of Department of Geology, University of Saskatchewan).

explosion, perhaps connected with the volcanism that accompanied the rifting of Gondwanaland, could do all that a cometary impact would do, and even provide the iridium.

Perhaps the Cretaceous did end with something like this – an impact or an explosion. The result would be a sudden catastrophic climate change as dust or smoke filled the upper atmosphere, followed over the next few months or years by the death of almost every large living thing on Earth. Of the higher animals, a few marsupials and some of the placentals survived. The tuatara, turtles, crocodiles and birds continued, but the dinosaurs that existed at the end of the Cretaceous died.

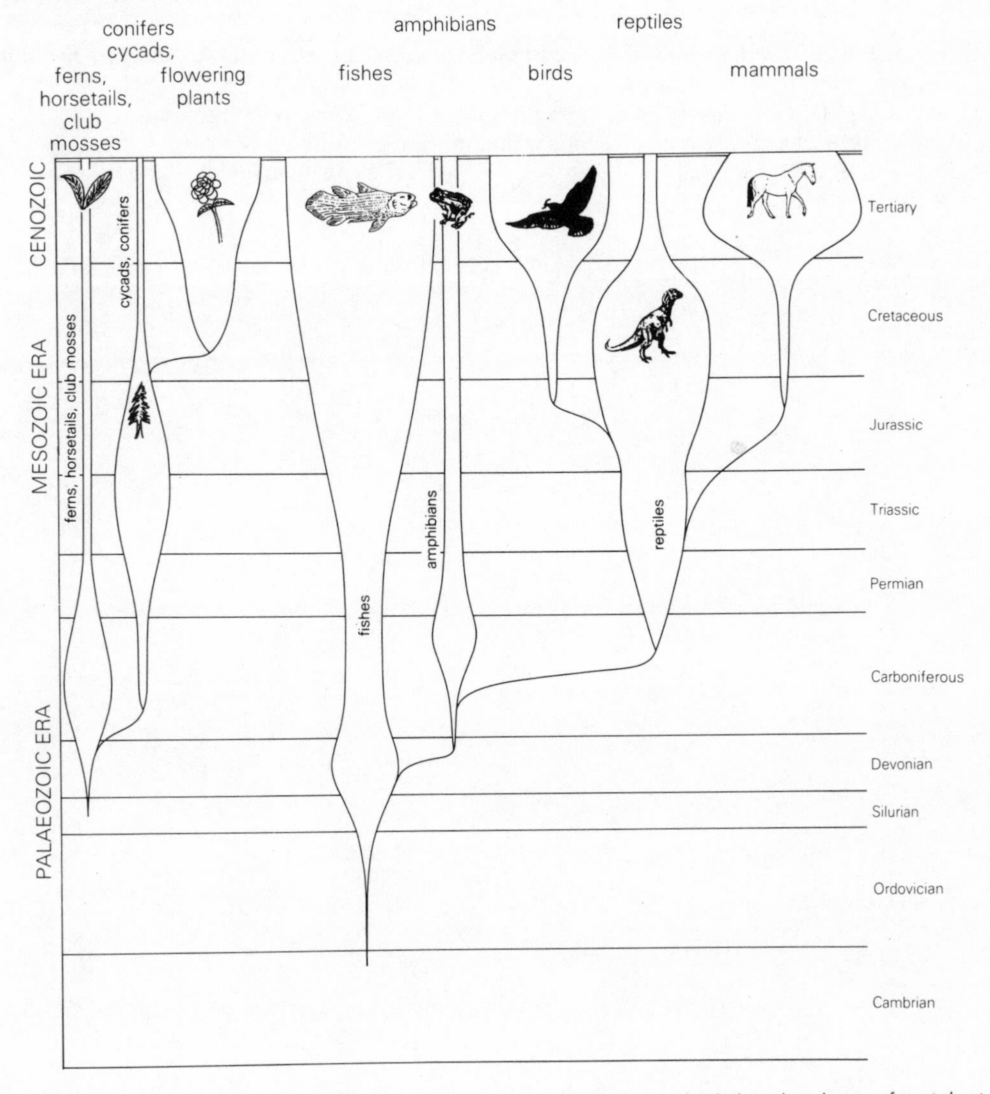

Figure 10.9 A simplified evolutionary chart showing the distribution and relative abundance of vertebrate animals and land plants through time (courtesy of Department of Geology, University of Saskatchewan).

Will we ever know for certain what happened? – perhaps not, unless we can find some truly undisturbed record that allows us to pin down exactly the length of time over which the change occurred, whether a million years or ten years. Whatever the cause, the ecological balance of the whole planet changed, especially in the seas.

'It is a poor sort of memory that only works backwards' said Alice. The message from the end of the Cretaceous is sombre and terrifying.

FURTHER READING

Bakker, R.T. 1986. *The dinosaur heresies: new theories unlocking the mystery of the dinosaurs and their extinction*. New York: W. Morrow.

Carroll, R.J. 1987. *Vertebrate palaeontology and evolution*. New York: W.H. Freeman.

Charig, A. 1979. *A new look at the dinosaurs*. London: Heinemann.

Radinsky, L.B. 1987. *The evolution of vertebrate design*. Chicago: University of Chicago Press.

PART 4

The modern world

11
The new world

A WORLD IN TURMOIL

There is a rainbow after a flood, and then the promise of new growth. The Cretaceous biosphere, which had developed over so long a time and had been so stable, was replaced by the beginnings of a new ecosystem that, over the tens of millions of years, slowly evolved into our familiar world of sheep and cows, lions and zebras, elephants, whales and people. A new world, diverse and complex, was built on the ruins of the old.

At the very beginning of the Tertiary, the planet seems to have been depopulated, or at least to have lost much of its diversity. The extraordinary variety of terrestrial life was gone; even wider extinctions had taken place among the small organisms in the sea. Was the ground covered with corpses for a year or two and the seas polluted under a grey sky, after a major meteorite impact or volcanic explosion? Or did the collapse take generations as a result of a spectacularly but briefly successful new species or disease? We do not know, but what remained seems to have been a bleak, impoverished planet.

Immediately at the end of the Cretaceous the climate seems to have changed sharply, especially in the part of west-central North America where the geological record is good. At this time at least four ecological regions can be distinguished, each with distinct plant populations. One region covered eastern Siberia and western North America, another Europe and eastern North America; the other regions were in the Southern Hemisphere. Immediately after the event that ended the Cretaceous, there was major change in the flora of the Siberian–American region, some alteration in Europe–America, but little effect in the south.

The most dramatic (or perhaps just the best studied) changes are seen in the rocks of the western interior of North America and in Japan where the previous vegetation, dominated by flowering plants, was replaced by ferns, the inhabitants of a miry swamp. In the western interior of North America, towards the Pole where the winter night was long, before the catastrophe the plants were broad-leaved deciduous trees (nowadays, we find surprising the notion of deciduous trees near the North Pole). After the catastrophe, rainfall in the western interior increased, and about half the earlier plant species were wiped out. One possible explanation for the changes, which in this region broadly favoured deciduous over evergreen plants, is that there was a brief but disastrous temperature dip. Possibly the temperature dropped for a couple of months, with the temperature close to freezing, (not too cold, because the crocodiles survived). This would have wiped out many of the large animals. At sea, however, where temperature is more stable, the record is of global disaster: there was widespread extinction of the plankton. Whole families of species were wiped out, and in some groups of plankton only one or two species survived. A simple temperature drop could not have done this.

Surviving in this emptiness were, in the seas, the remaining plankton species, assorted invertebrates, sharks and fish. On land some crocodiles (did they huddle in swamps and eat carrion?), snakes, lizards and tortoises, and some species of mammals continued. Of the dinosaurs, only the birds were left. Beneath all the animal ecosystem remained the backbone of life, the bacteria, but even they must have adjusted as the ecological niches changed and the vegetation altered.

How did the biosphere recover? To understand the fossil record of the recovery, it is necessary to study the forces that shaped the rebuilding. These forces include the subtle interplay between climate and life, the brute force of continental drift, and the complex controls on the biosphere's carbon budget.

THE SPLIT OF THE CONTINENTS

What did the planet look like physically? The New Zealand short-tailed bat, which as a species is today very near extinction, provides an introduction. There is molecular evidence that this bat is related to bats in Central America. How did a Costa Rican bat cross the Pacific? Plate tectonic reconstructions give the answer. The reconstructions have been put together by reading the record of spreading that is locked in the magnetization of the ocean floor. In the early Tertiary the Americas, Antarctica and Australasia were connected. The bats, though they can fly far over sea, were able to spread mostly over land from America to New Zealand.

The distribution of the continents in the Tertiary has become one of the most powerful factors controlling the pattern of life on the modern planet. The separate continental arks carried with them distinct groups of plants and animals (and fleas), which give excellent evidence of continental drift in their distribution of species. Each ark developed a complex ecosystem. From time to time some species crossed between these systems, such as between the Americas, or between Asia and North America, but other fragments of continent such as Australia, New Zealand, Madagascar or the Seychelles became lost at sea.

The breakup of Gondwana began in earnest during dinosaur time in the Jurassic and Cretaceous. Enormous volumes of lava were erupted along the rifting margins. In southern Africa alone, over five million square kilometres (approx. two million square miles) are covered in Karoo lava flows and intrusions (the Victoria Falls are built in them); in South America vast areas of Argentina and Brazil were covered (the Iguazu Falls). Antarctica and Australia too have extensive sequences of lava, while in India the Deccan Traps are a similar gigantic suite of flows. By the beginning of the Tertiary, the mid-Indian ocean ridge was close to Bombay, a circumstance which would have greatly enlivened the backdrop to proceedings in that capital of the world's movie industry. In the North Atlantic, by contrast, things were slower to get started, but eventually as the ocean opened Greenland and Glasgow were having a torrid time of it, and unspeakable volcanic events were happening in the malt whisky islands of Scotland, not to mention construction of the Giant's Causeway in Ireland.

Older geological texts have a notion of great mountain-building events and great earth movements. These do indeed occur, but for the most part are not a particular nuisance to life – the old textbook descriptions of 'paroxysmal' events hides the between-earthquakes normality of San Francisco or Auckland. We see today a great mountain-building event in progress along a line of deformation from Greece to China. Yet Greece is a splendidly habitable place (apart from the occasional earthquake or eruption) as is China. Likewise, active volcanic areas such as Japan or New Zealand may have their dangers but are not uninhabitable. The beginning of the Tertiary was probably like this, with major orogenic events in progress at various places, and massive volcanic eruptions in others. Yet possibly there was a difference in scale compared with today – the huge floods of lava that accompanied the splitting of Gondwana from the Jurassic to mid-Tertiary do not have a modern equivalent, except perhaps in Iceland and in the Danakil depression in Djibouti in northeast Africa. The Gondwanan lava flows must have been much more dramatic than anything in human history.

The Atlantic opened slowly, first in the south and centre and then, much later, the gap between Newfoundland and Ireland widened and Canada split from Britain. Interestingly, the equatorial Atlantic widened by slipping sideways and what today is Brazil's north-east coast slid along the coast of Nigeria, Ghana and Sierra Leone. For much of the early Tertiary this gap was narrow and may have had land bridges on volcanic ridges. In the north, Canada first began to split from Greenland. Then the line of opening shifted, and the North Atlantic opened out between Greenland and Scotland. Iceland was born.

In the Indian ocean, the Indian fragment was speeding northwards (geologically speaking). It finally came close to Asia about 40 million years ago. To the south, Australia and Antarctica initially formed a single large continent which had split from Africa in the Cretaceous. Antarctica lay across the South Pole, as now, but it had forests and was not iced over. New Zealand was probably isolated, but not greatly, from Australia–Antarctica. On the other side of the Pole, South America remained in contact with Antarctica and Australasia, and also maintained a tenuous link with North America.

THE CHANGING CLIMATE

The Tertiary was a period quite unlike the Cretaceous. Much of the previous period was a time of stable warm climate and wide shallow inland seas. In contrast, the history of climate in the Tertiary is a story of complex shifts, warm episodes and cold spells, but in general a cooling trend, eventually turning to the glaciation in which we now live (although we happen to occupy a brief warm spell in the Ice Ages). This general cooling of the climate may have been caused by several factors. On a global scale, the drift of the large land masses would have influenced climate significantly, and may have changed the way in which heat was transferred from the equator to the poles. Simultaneously, the greenhouse contribution from carbon dioxide may have declined, cooling the planet. More recently, the complex detail of the shifts of temperature in the Ice Ages seems to be closely connected with the subtle changes in the Earth's orbit and the tilt of its axis of rotation, which provide an excellent explanation of the cyclical ups and downs of temperature in the past half million or so years. These changes are known as Milankovitch effects.

Climatic change is a poorly understood business, as the workings of the planetary atmosphere and oceans are so complex. The temperature of the Earth is subtly set, by its insulating and reflective blanket of clouds, by its surface colour and by the greenhouse gases, especially carbon dioxide, CO_2, and methane, CH_4. Clouds reflect heat, as do ice caps; dark surfaces absorb it. Across the modern planet, the mass effect of biological organisms conspires to set the temperature, by generating clouds across the rainforests, by adding to the air nuclei that help the condensation of water vapour, or by changing the colour of the surface. There is no simple, direct link between the surface temperature and the amount of radiation coming from the Sun. If temperature falls, biological growth may decline, with the effect of slowing or correcting the fall, and vice versa. At the end of the Cretaceous, the temperature was probably slowly dropping, but the plants and animals were probably changing with the temperature, accommodating to it, perhaps stabilizing it. There seems to be no reason why the dinosaurs could not have withstood a slow evolutionary change to a much cooler climate. In contrast, a series of strong climatic fluctuations, or a single sudden, short, sharp drop in temperature would have finished them off.

The climate determines the habitat of life, but to some extent the reverse is also true. If massive changes occur in the pattern of life, then the climate must change. Forests affect climate, climate affects forests. Animals inhabit forests: if the animals die, then often the forests will die too. If the forests die, then the climate and temperature will change.

There is a tale from Mauritius that illustrates the impact of animals on vegetation. The island has a species of fine trees, mostly over three centuries old. They bear seeds, but the seeds have a very hard shell and they rarely sprout: young trees are not growing and, in consequence, the trees have become endangered and seemed doomed. Because trees interact with the atmosphere, in a minor way (Mauritius is a small place) the local environment must have changed too as the trees became rarer. Enter Scientific Genius, who made the brilliant association with the dodo, collected some turkeys and force-fed them with the seeds. The seeds, of course, were excreted by the birds much softened, and sprouted in the dung. The Mauritian forest service, also displaying genius, succeeded in a different way, by abrading the seeds before planting them. The tree species has been saved. Did the dodo eat the seeds and was there a symbiosis between the two? Probably, though we cannot be certain.

This little tale demonstrates a greater truth, that most plants and animals are intimately interdependent. A thriving tomato plant was recently reported by the tracks of Kings Cross railway station, London: a stony desert except for the discharge of train toilets rich in manure. An excellent place for a pip to grow is in a dungheap, perhaps miles away from the parent plant, so in some circumstances it can be advantageous for a plant's survival if its fruits are attractive to the eater. Bearing edible fruit with seeds inside can be a successful evolutionary strategy: the ecosystem is interdependent. When the dinosaurs went, so did many of the plants on which they depended and which in turn depended on them. And, quite possibly, as the vegetation changed, the climate changed too as a result.

It is not only physical changes that cause collapse. There is a small island off New Zealand called Big South Cape Island, where in 1964 a disastrous extinction of animal species took place. Some ship rats arrived in 1962. By 1964 their population exploded – the maths goes 2, 8, 32, 128, 512, 2048, 8192, 32768, 131072, 524288, 2097152, assuming four offspring to each rat. Four species of birds immediately became extinct, others were threatened or disappeared later, and a subspecies of the New Zealand short-tailed bat went too. The result of all this, no doubt, will be a complete change in the island's ecosystem, say, two centuries hence. Plants and animals will be quite altered. The local microclimate will be different too, depending on how many trees survive.

Perhaps the event that terminated the Cretaceous was catastrophic, a sudden killing of plants, under a dark sky of soot or volcanic dust, or cometary impact dust. Or perhaps it was a slower extinction, as the plants failed to reproduce when the ecological balance collapsed. Or possibly – we do not know – the apparent collapse was simply the culmination of a gradual series of extinctions.

THE CARBON BUDGET

Over the long term – hundreds of millions or billions of years – the carbon content of the Earth's surface layers is governed by the balance between, on the one hand, the degassing of carbon from the planet's interior by volcanic eruptions and, on the other, the return of carbon to the mantle down subduction zones. The gigantic igneous outbursts in the Jurassic and Cretaceous must have introduced huge amounts of carbon dioxide into the air; part of this carbon was captured biologically. For example, the Cretaceous geological record includes formations in western Canada and western Siberia so rich in organic matter that they contain thousands of times more carbon than the entire modern biosphere and atmosphere. The chalk formations of England and western Europe are calcium carbonate, made of the bodies of countless small marine organisms called coccoliths that have captured carbon dioxide. Most probably the Cretaceous air was rich in the CO_2 originally vented by the extensive lava flows and from volcanoes and then captured in oil sands, chalk and limestone.

Igneous outbursts continued in the early Tertiary, but then became less frequent and smaller; the source of carbon dioxide diminished. New sea floor opened, and the mid-ocean ridges in the deepening oceans were net sinks of carbon, not sources, as hydrothermal systems precipitated calcite. On the margins of the new continents, thick sediment accumulated, rich in organic matter, removing carbon from the biosphere. Today, in large areas of the world's ocean floors, geophysicists can map a seismic layer known by them as the bottom-simulating reflector. This marks a layer of methane gas held in the sediment, captured in a salt-like material known as gas hydrate. Gas hydrates on the ocean floor and in the Arctic may contain anywhere between 1–1000 times as much carbon as the modern biosphere. The carbon is derived from organic matter in the sediment, originally extracted out of the atmosphere.

Most probably, the sinks of carbon during the Tertiary were larger than the sources, and the standing crop of carbon in the air and in living things has been slowly depleted by a steady, small drain on its resources. CO_2 levels in the air have probably fallen during the Tertiary, cooling the planet, although occasional volcanic outbursts, such as the eruption of floods of lava in Ethiopia 25 million years ago, must have recharged the resources somewhat. Since about 10 million years ago there have been no massive outbursts. The geometry of continental drift may also have abstracted carbon from the system. This is because the burial of carbon in sediment in the new Tertiary oceans has not been equally offset by the excavation and release of carbon from metamorphism and volcanism in mountain belts. Overall, the biospheric system has probably had to cope with a major loss of carbon since the beginning of the Tertiary.

All this – the catastrophe and its after-effects, the physical changes in the Earth's face as the continents moved about, the changing climate and atmosphere – created new challenges and new opportunities for the vertebrates that survived on land. The most successful response to the opportunities came from the mammals, including our own ancestors.

THE SPREAD OF THE MAMMALS

The first modern mammals probably evolved about 200 million years ago, from creatures of the same general type as Mr Bain's African monsters. Bain's monsters were herbivores; their cousins, from which modern mammals are descended, were smaller carnivorous species. The earliest mammals were small warm-blooded creatures that probably ate insects. Today there are three main divisions – the monotremes, the marsupials, and the placentals.

The monotremes

The living monotremes are the duck-billed platypus and the echidna, which lay eggs but exude milk for their young. They can be viewed as a continuation of a branch of the mammal-like reptiles, still living today in specialized settings. One platypus-like beast is known from the early Cretaceous, also in Australia. The platypus line is old, but it is not necessarily primitive – the platypus, like almost all organisms, is a well adapted and highly evolved beast, finely suited to its setting (it even finds its food by electrolocation, which is much better than depending on sight in muddy billabongs at night). In its own way, the anteating echidna which lives in Australia and New Guinea is also a sophisticated beast, well adapted to its environment.

The marsupials

The second major division of mammals is the marsupial group. The split between the marsupials, which carry their young in pouches, and placentals, which carry babies internally, probably occurred early in the Cretaceous. By the beginning of the Tertiary, these two groups were very distinct. Fossil and living marsupials seem to be mainly restricted to the Americas, Antarctica and Australia – historically there were few in Asia or Africa, although remains are known from Africa and a marsupial fossil is known from the Asiatic USSR. In Europe they died out about 20 million years ago, only recently being reintroduced – the Tasmanian wallaby, which has formed successful colonies in

Figure 11.1 The duck-billed platypus, *Ornithorhynchus*, a monotreme. This egg-laying mammal can find food by sensitive electrolocation organs. About 70 cm–1 m (2–3 ft) (courtesy of Department of Geology, University of Saskatchewan).

England, is now regarded by some as a wild British animal (nearly all British animals, including people, include in their genetic heritage ancestors who were recent, post-glacial, immigrants). Quite possibly, the marsupials began in North America and spread to South America across the tenuous Caribbean bridge, then via Tierra del Fuego and Antarctica to Australia which they reached in the late Oligocene.

Marsupials are superbly diversified, and have produced forms analogous to most of the more familiar placental animals. One of the most fascinating of these is the Tasmanian wolf, which closely parallels its northern canine equivalents, and which may be extinct (although sightings are reported: there is still hope that it is alive). Other marsupials in Australia, now extinct, included giant kangaroos browsing trees like giraffes, a rhinoceros-like beast, and even a marsupial 'lion' which may or may not have been a nasty predator. Other groups in Australia flourished too: for instance, there is a fine fossil snake with the delightful name *Montypythonoides*. One fascinating predator was reminiscent of the days of the dinosaurs. This was a huge monitor, or lizard, up to 5 m (15–20 ft) long and weighing up to 2 tonnes. Perhaps it ate some of the giant kangaroos and wombats. The modern Komodo Dragon, also a monitor lizard, found in Indonesia is a relative of this lizard. One suggestion is that it dined on pygmy elephants before man conveniently brought in domestic animals for the dragon to eat, about five thousand years ago.

Many of the Australian animals died out very recently under the impact of aboriginal and western man, the dingo and cats, but not all is disaster for the Australasian marsupials. Apart from their colonization of England, marsupials have been successful in occupying New Zealand recently, where the roads are littered with the pelts of that remarkably successful recent immigrant, the Australian possum, which is busy devastating the Kiwi forests. A species of wallaby that is endangered in Australia seems to be firmly established on an island in Auckland harbour, where it was introduced many years ago by a romantically minded governor and premier. In North America, recently, there were reports of a feral kangaroo attempting to overwinter in a Saskatchewan grain field, but this animal was perhaps slightly over-optimistic in its anticipation of the greenhouse effect.

In contrast, the marsupials of the Western Hemisphere have had a much less successful history in the past few million years. In North America they are today only a minor part of the fauna. The South American marsupials, however, were isolated for much of the Tertiary and flourished. They developed an array of skills, both as carnivores and herbivores, with many species. When the two continents again became linked and the Panama bridge reopened the northern animals arrived and they may have helped to reduce the diversity of the marsupials, many of which became extinct. The American opossum was a marked exception to this story of extinction. It has occupied North America and seems securely established. Antarctica's marsupials had a sadder end – the climate grew colder and even possums could not survive the ice.

The placentals

Some years ago one of my prospector acquaintances in Zimbabwe was disturbed from his normal business when a panic arose underground in his smallholding. The place erupted with men running away from what they called a 'tiger'. Now Africa is not noted for its tiger-shoots, and the miner bravely went below. Suddenly he saw it too in the dim light – a striped carnivore, and he too fell victim to fear and panic.

The animal was an aardwolf, a small beast with simple teeth, mostly living on insects and grubs. It fluffs up its hair and looks bigger when frightened. The aardwolf is a shy, retiring animal, not obvious on the landscape of Africa. It was from animals with these habits – small generalized insectivores that scuttled around in the darkness and the undergrowth – that the wide spread of the mammals began. The aardwolf is specialized in its own way, and its simple teeth are actually an efficient and well designed eating apparatus. The ancestors of most modern mammals may have had habits like the aardwolf, though the aardwolf is evolved, and genetically not especially close to them.

Most probably, the early ancestors of most mammals were similar to shrews, moles and the like. From these shrew-like ancestors have developed most of the modern mammals, a diverse population of eaters and eaten, walkers, swimmers and flyers. This group, the largest division of modern mammals, is known as the **placentals**.

Placental mammals have a great asset, which has helped them to populate the globe so widely. This is the placenta (or afterbirth), which gives them their common name. In both marsupials and placentals, the egg is very small and the yolk is poor. In marsupials the egg develops for a few days and then hatches as a tiny neonate that struggles to its mother's pouch, there to suckle. In placentals, the fertilized egg divides and two distinct cell masses develop. One becomes the embryo, the other develops into a complex chemical factory that includes a protective barrier between the embryo and the mother. This barrier prevents rejection of the embryo by the mother's body, and it also secretes hormones that signal to the mother that implantation has taken place. Once the barrier is in place, the embryo can develop securely, as long as the mother is alive and well.

Figure 11.2 An Oligocene cat (photo courtesy of Royal Tyrrell Museum of Palaeontology/Alberta Culture and Multiculturalism, Canada).

Their method of reproduction seems to have given the placentals major advantages over the marsupials. Although, in terms of energy, the cost of giving birth is small in marsupials, and the young can be jettisoned in time of danger, the total cost of creating an offspring is higher because of the metabolic cost of lengthy lactation. Among animals of similar size and environmental occupation, placentals appear to be able to reproduce more rapidly.

THE EXPLOSION OF THE PLACENTALS

Marsupials and placentals probably diverged from each other during the Cretaceous, perhaps sharing a common ancestral line in the early Cretaceous. By the end of the Cretaceous a diverse placental fauna was established. Fossils from the western parts of Canada and the United States record this community and also record its descendants that survived into the Tertiary. During the latest Cretaceous and early Tertiary there was wide diversification of the mammals. From the original shrew-like animals there developed several distinct groups. At the end of the Cretaceous, the early placentals included the shrew-like animals, other insectivores, an ancestral primate and a variety of condylarths, which were herbivores. These groups soon split further. From the condylarths are descended animals such as the cow, the camel, the hippo and the elephant, as well as the aardvark and perhaps the whales. The carnivores, primates and bats may have diverged earlier and the rabbits and rodents may also trace their ancestry back to Upper Cretaceous animals. Pangolins, or scaly anteaters, too, may have become a distinct group in the Cretaceous.

By the early Eocene, nearly all the modern types of placentals were distinct. The herbivores had diversified to occupy the northern continents and Africa, and were common in South America. The early condylarths had teeth that show only limited specialization for feeding on plant material. Later, their descendants developed large molars, with increased surface areas to grind and crush plant materials. Selective feeding developed, and individual lines of animals evolved to specialize on particular plants or types of vegetation. In the Miocene, the spread of the grasses was associated with the development in many herbivores, the ruminants, of complex digestive tracts to extract nutrients from plant material. A cow chewing the cud is an example. Others, the horses, pass food relatively rapidly through the digestive tract, and can eat food that is relatively unnutritious and fibrous. Between 15–18 million years ago, in the early Miocene, at least 19 new species of horse-like animal appeared. A mixed population of herbivores, such as zebra, buffalo and antelope in Africa, is able to make full use of a range of foods. The horse-like animals can feed on and control plants that are inaccessible to the other herbivores. Other animals eat tree leaves. Manure then helps to recycle the nutrients. In this way a complex community can be sustained, with a much higher biomass than if only one species of herbivore were present – African cattle farms have far less meat on the hoof than equivalent land occupied by wildlife.

The spread of the herbivores was matched by the spread of the carnivores, but the change from eating insects to eating the flesh of other mammals seems

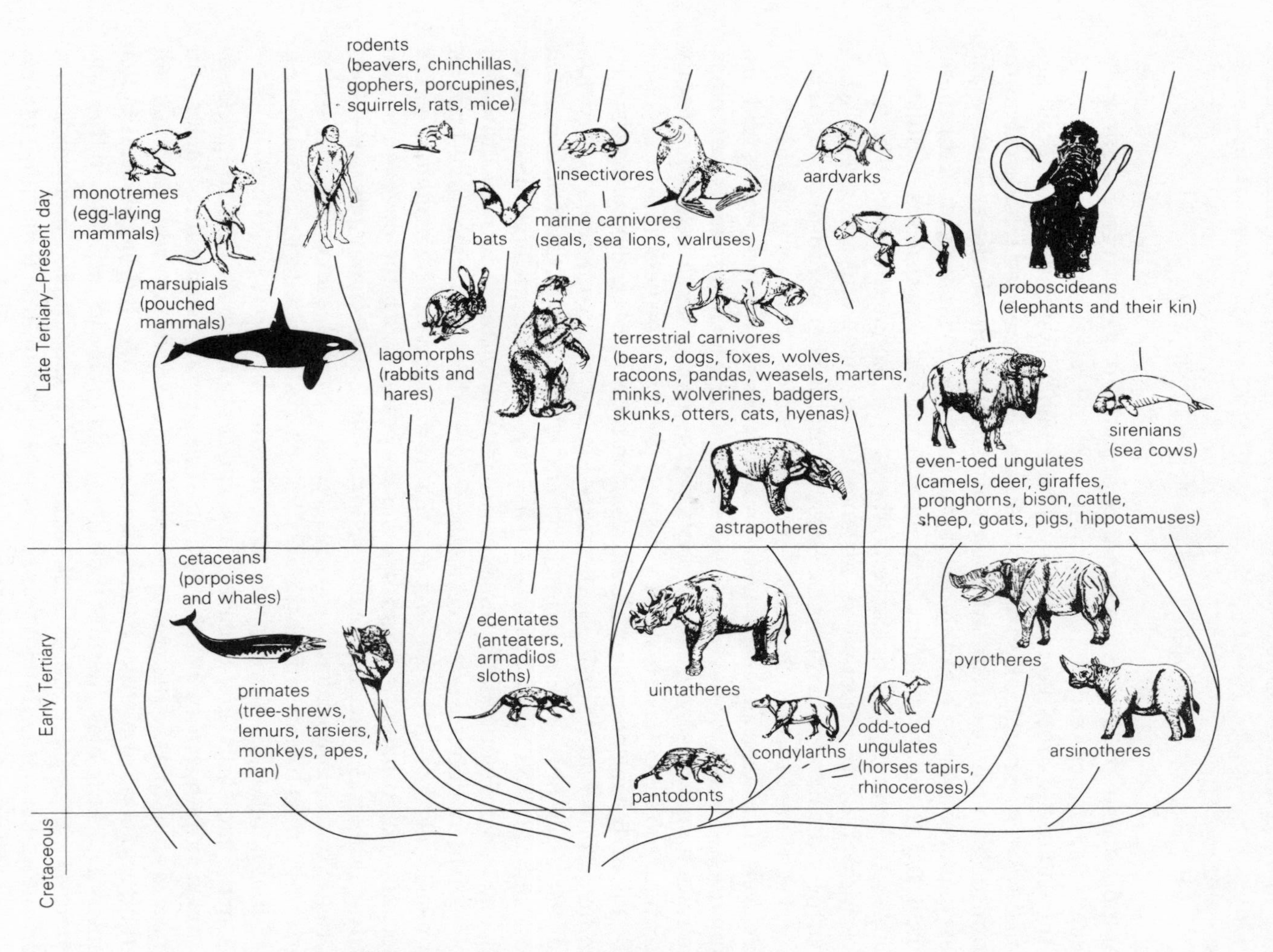

Figure 11.3 The evolution of the mammals in the Tertiary.

to have been relatively slow. Only towards the end of the Palaeocene did moderately sized meat-eating animals begin to fill the gaps left after the end of the dinosaur predators. Two early groups of carnivorous placentals appeared. One, the creodonts, flourished in the Eocene and survived until the Pliocene but is now extinct. Creodonts ranged up to the size of large bears, and some seem to have had large brains, but were eventually replaced by the other group, the modern carnivores or Carnivora. Carnivora include hyenas, cats, dogs, bears, weasels – our familiar modern predators. Some went to sea. One group of these is now represented by the seals, another by the sea-lions and walruses.

The whales and dolphins, in contrast, may be descended from condylarths which took to the sea in the early Tertiary. In the Eocene, the earliest whales may still have spent some of their lives on land. Later, the land was abandoned entirely, and the animals even became able to give birth at sea. Some whales are herbivores, living on plankton. Others eat fish, squid or fellow whales. The blue whales are bigger than the big plant-eating dinosaurs, the sperm whale more ferocious and probably much more intelligent than *Tyrannosaurus*.

Perhaps the most successful of all mammals, if success is judged by sheer numbers, have been the rodents. Today they include rats, mice, and also animals such as marmots and beavers. The smaller rodents reproduce extremely rapidly – a hamster can become pregnant a few weeks after its own birth. Large rodents have existed too – some North American beavers, now extinct, have been the size of bears. On land, and at sea, the mammals have produced extraordinary diversity.

The spread, or 'radiation', of the mammals has mostly occurred during the last sixty million years or so; this has been less than half the time the dinosaurs existed. Natural selection has driven the evolution of mammalian skills. Hunters and hunted live in quasi-equilibrium: if a hunter is too successful, the prey becomes extinct, so the successful hunter dies too. The ecology of a community is in a constant flux around equilibrium, a dynamic spiral never quite attaining stasis, rarely drifting far from it. Slow and stupid prey are killed, inefficient hunters die. Only rarely does a herbivore die peacefully, unless, like elephants, they are so huge as to be immune from predators. Carnivores can perish too, both individually and as a species.

Most of the early (Triassic to late Cretaceous) mammals seem to have been carnivores, though their prey in many cases may have been insects. During the evolutionary radiation, some lines remained carnivorous, but others became herbivores, exploiting the changing vegetation and helping to create that change as plants and animals co-evolved. This transition from carnivore to herbivore is a common theme of the history of life: the derived herbivores eventually seem to become extinct to be replaced by other herbivores derived from a continuing carnivore line.

Plants and animals co-evolve. If the food changes, the eater must change, and often the opposite is also true. The evolution of the placental families, especially the herbivores such as the ruminants and horses, proceeded in parallel with major changes in the vegetation.

THE C_3 AND C_4 PLANTS

Tertiary and modern plants have three different ways of making carbon compounds from carbon dioxide. These ways are named according to the carbon compound they initially produce: C_3 (phosphoglycerate, a three-carbon compound), C_4 (oxaloacetate, a four-carbon compound) or CAM (crassulacean acid metabolism). C_3 plants are in the majority, especially trees and bushes. They thrive in air with high CO_2 content. In contrast, C_4 plants do well in hot environments when atmospheric CO_2 is low. Examples of C_4 plants include maize, sugar cane and some other grasses. CAM plants are succulents, adapted to desert conditions.

During the Tertiary the evolution and spread of C_4 grasses were probably favoured at a time of low atmospheric CO_2. The competition between C_3 trees and C_4 grasses may aid the fine-scale global management of the greenhouse. C_4 plants do well in the tropics and subtropics during ice ages which are periods of low CO_2, C_3 plants during warmer times when atmospheric CO_2 and rainfall is more abundant. If the Earth becomes cold, C_3 trees do not prosper and less carbon is extracted from the air; on the other hand, if volcanoes release

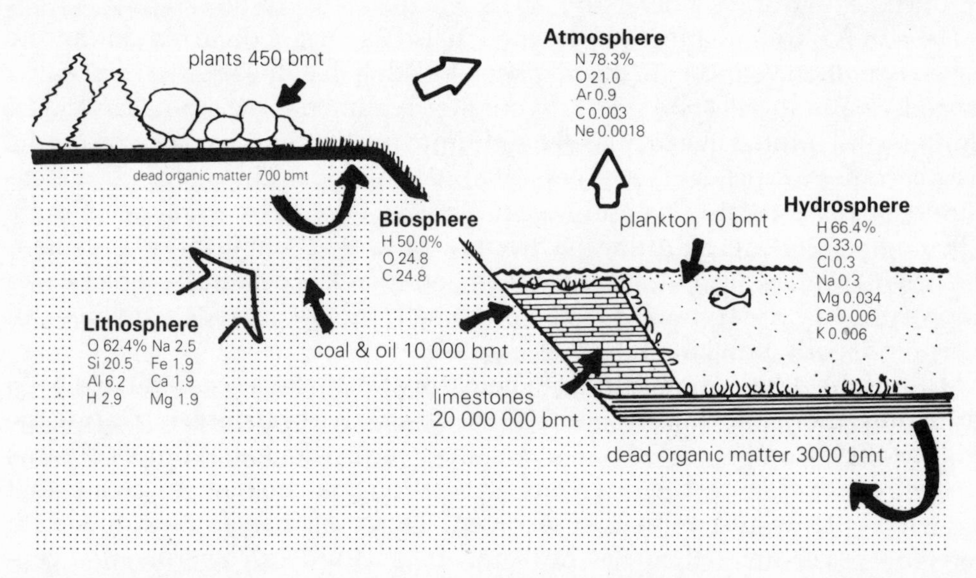

Figure 11.4 The cycling of elements in the biosphere. Figures refer to the concentration of carbon in the global surface system, expressed in billion metric tons (bmt), or to per cent composition of systems. Most carbon is in limestone, in long-term storage, and in fossil fuels. In medium-term storage, carbon is held in sea-floor sediment and in soils. In the active biosphere, most carbon is on land; little is held in sea life (from Copper, P. 1988. *Geoscience Canada* **15**, 199–208).

CO_2 to the air, the C_3 plants will flourish, withdrawing carbon from the biosphere.

In the Miocene, the grasses spread across the continents. The plants helped to shape the evolving animal communities. In Australia, the grasses and spinifex plants co-evolved with the varieties of kangaroo and other large marsupials; in the Old World the equivalent to these herds of hoppers is the complex community of antelope, horse-like animals and cow-like animals in Africa and Asia. The 'dawn horse' was about the size of a fox – with the spread of the grasses across the continents, the later horses became larger and faster. By 15 million years ago, about 16 different species of horse grazed the plains of North America. Rhinos and tapirs appeared and grew large, and animals with trunks also spread across the continents. The other herbivores, widespread and numerous, now include pigs, hippos, deer, antelope, sheep, goats, cattle and camels. Ruminants, such as cows, can effectively digest grass and leaves, which are far less nourishing than the fruits that their early ancestors may have eaten. Cattle and sheep are, today, perhaps the most successful of the herbivores in that they have persuaded their chief predator to feed them, house them and transport them across the globe.

THE BIRDS

The Tertiary was the time of mammals, but in the birds one line of dinosaur-like animals did survive, as a small reminder of the Cretaceous. Mammals are poorly adapted for flying in the air, having to carry their embryos with them, either in a womb or a pouch, although the bats have produced fine flyers, and some marsupials are skilled gliders. In contrast, birds do not need to fly carrying their eggs, and can abandon them if necessary when danger threatens, with only minor loss. Furthermore, the birds colonized the air long before the radiation of the mammals, and the competition from birds would have been very difficult for prospective mammalian aviators: only the bats managed to find a special niche, the night, aided by their sonar navigation skills which exploit the assets of the mammalian brain and ear.

Having survived into the Tertiary, it is perhaps surprising that birds did not to some extent displace mammals from the land or occupy the sea. Presumably the loss of grasping forelimbs was an insurmountable obstacle, but for many purposes the bird's digestive system, which (like that of the dinosaurs) includes a gizzard, would have been a significant advantage: less chewing is needed, and teeth can be simpler. One bird, the South American hoatzin (already mentioned above for the teeth and wing claws in its young), has even developed a digestive tract which is analogous to that of ruminant mammals. In the rainforest, this digestive system has allowed it to become the avian equivalent of a cow.

On land, some of the most successful birds are those that are large and flightless, like the ostrich, which survives in an environment occupied by predators as fierce as the lion and leopard. At sea, penguins in the Southern Hemisphere and the great auk in the Atlantic (now extinct) developed into superb swimmers. Unfortunately, they make fine meals for sea-lions: the attempt by the birds to colonize the sea has been limited by the presence of the mammals.

On some islands the birds were free of the competition from mammals. The largest of these areas was the New Zealand landmass, where the birds had a chance to create their own ecosystem. A magnificent array of birds developed in a setting free from mammalian competition. The flightless birds of New Zealand included the many species of moa (some much larger than ostriches), the takahe (similar to a dodo, and once thought extinct but still tenuously surviving – there are a few hundred left under heavy threat from introduced wapiti and red deer), the species of kiwi, the giant kakapo parrots (a few females, on Stewart and Little Barrier Islands, and rather more males, on the mainland, left) and penguins. Others include an array of sea-birds, and innumerable small birds, such as the Stephen's Island wren (once a cohabitant of the tuatara, but only known from the handful of specimens brought in by the lighthouse keeper's cat, which extinguished the species).

X-ray evidence of bone development suggests that many of the larger flightless birds – rhea in South America, ostrich in Africa, cassowary in New Guinea, emu in Australia and moa and kiwi in New Zealand – seem to be descended from a common stock. From this stock, the ancestors of the moas seem to have crossed early into New Zealand and flourished.

New Zealand had no mammals except its two species of bat (placentals), one from South America and one from Australia. Perhaps New Zealand was so isolated or inundated that no marsupials migrated there until humanity brought the possum and the wallaby. Somehow the tuatara survived, as did an assortment of frogs and geckos, but the islands were dominated by birds, including the ancestors of the moas and kiwis (plus their fleas). Plants, too, crossed from South America and Australia. The most famous is the southern or false beech (the scientific name, *Nothofagus*, is a clever pun which can mean either, depending on how adventurously the English is transliterated back into Greek). This tree family occurs today in the southern tip of South America, Tasmania and New Zealand – testimony to continental drift.

Left to themselves, the birds of New Zealand developed an assembly of talents which matched those of the mammals in other continents. However, in the absence of agile ground-based predators, birds such as the kiwis lost the ability to fly, not needing it to escape, and adapted to a ground-based life. As Charles Darwin realized during his visit to New Zealand, the introduction of people, dogs and rats (which came during Maori colonization) and then cats, and, later, weasels, stoats, rabbits and so on (brought by Europeans), completely shifted the balance, driving many species towards extinction. A single dog recently killed a significant proportion of the remaining wild kiwis, after the authorities had lessened the degree of protection under which the birds lived.

Curiously, though, in the less competitive setting of New Zealand some animals that are failing elsewhere can thrive. For instance, the chamois, which is rare in Europe because of human activities, has colonized parts of New Zealand in such numbers that it is a pest. Sir George Crey was the early governor and premier who introduced wallabies to an offshore island: an early history of New Zealand predicted that 'when Australia shall lament the wholesale destruction of her unique fauna, the . . . survivors of the quaint marsupial order shall, perhaps be found on the isle'. Seventy years after the prediction, much of the surviving population of that species of wallaby was on that New Zealand island.

For the native flora and fauna of New Zealand, today the only temporarily secure fragment of that great Ark is in the few thousand immensely precious acres of Little Barrier Island. Should this island be lost also, humanity too will lose something valuable: a part of its soul. Of all our earthly heritage the fate of these few obscure acres is most properly seen as the touchstone of our common future.

Birds, monotremes and marsupials have done wonderful things, but they were also-rans in the competition to fill the gap left by the dinosaurs. The land birds have colonized the islands, especially New Zealand, and have had a major presence in Madagascar, Australasia and even, as ostriches, in the most competitive setting of all, the African–Asian landmass. Birds have retained their dominance of the air and they have explored the seas. In contrast, the monotremes persist in dark corners, and the marsupials have held one isolated continent, with a few surviving quietly in the Americas. In a sense the relationship between the marsupials and ourselves is comparable to the relationship between the successful dinosaurs and the less widespread mammals in the Cretaceous – the modern marsupials are a distantly related, less successful group than our relatives. The placental mammals are arguably the most successful line of animals ever. Only they have managed to produce a predator so fearsome that, were the dinosaur *Tyrannosaurus rex* magically to reappear today, it would promptly be put in a cage for the amusement of children.

FURTHER READING

Stahl, B.J. 1985. *Vertebrate history: problems in evolution*. New York: Dover.
Stanley, S.M. 1989. *Earth and life through time*, 2nd edn. New York: W.H. Freeman.

12
Humanity

THE EVOLUTION OF THE PRIMATES

In the fossil record of the Upper Cretaceous there are traces of the mammalian groups that later diversified in the Tertiary. One of the most tantalizing of these precursors is the fossil of a single tooth that appears to have belonged to a very primitive representative of the line of vertebrates called primates. Humans are primates. The animal that owned the tooth is appropriately called *Purgatorius*, after Purgatory Hill in the Hell Creek Formation.

The ancestral mammals, being related to the early five-toed reptiles of so many years before, had five digits on their feet, which is why we now count our numbers according to base ten rather than the mathematically more flexible twelve: in evolution seemingly small early choices have long-lasting consequences. The early ancestors of the primates probably had five toes on each foot, together with generalized teeth not specifically designed for eating flesh or plant, but for grabbing whatever came to hand, mostly insects.

The tree shrews, which today live in the East Indies, are probably similar to our ancestral stock. A distantly related primitive line is that of the lemurs, which are today found mainly in Madagascar (where they are now close to extinction from forest destruction). Most lemurs live in trees and move around at night. They have good noses, and eyes which are more to the side of the head than forward-looking. Another primitive primate is *Tarsius*, which lives in the East Indies and the Philippines. This little beast has a small nose and two eyes turned forwards: it has what we would call a face, not an animal's snout. One of the distinctive developments in primate evolution was the adaptation to living in trees and, as a result, the dependence on seeing, not smelling. Most non-primate animals have two separate eyes, each of which covers a distinct territory: there is little overlap. In contrast, the majority of primates have overlapping vision from their eyes, giving stereoscopic coordination. They also have good colour vision.

Good sight and depth perception were important to our ancestors, who needed to be able to grasp fruit and to distinguish between ripe and unripe

Figure 12.1 A pygmy shrew (courtesy of Department of Geology, University of Saskatchewan).

colours. By looking at the same thing from two slightly different points of view, using two eyes, we see depth. Stereo vision demands a flat face, and two eyes looking forwards. In contrast, for instance, to the chameleon, we cannot look out of the side of our head. For most animals, good sideways vision is important, for attack (seeing prey) and defence. Primates, however, have less need of sideways vision because they have a social structure, with good communication between individuals: one animal looks forwards, while others in the group look sideways for good food, or to look out for danger.

Other animals, lacking a social structure, need wide fields of vision and so have little overlap in what they can see from their eyes. Nevertheless, they can still judge distance, though not as well as humans can. They judge distance by parallax – moving the head slightly to and fro, so that the relative positions of two distant objects can be judged. To understand parallax, hold up two fingers, one at a distance of about 15 cm (six inches) and the other at arm's length, in line away from the eye. Moving the head moves the near finger either to the left or right of the far finger. The brain turns this movement into depth. The judging of depth with one eye depends on memory. A scene is 'photographed'; a slight movement of the eye gives a new image. Superimposing the new image and the old mental photograph gives a perception of depth. Done intuitively and automatically, this method gives excellent depth perception – I myself have only one eye, but I can drive a car and attempt, badly, to play tennis – but at distances shorter than about 10 m (30 ft) (e.g. for gathering fruit), or for fast-moving objects it is inferior to the two eye method. For an animal that is precariously perched on a tree branch, the two eye method is much safer. More

Figure 12.2 *Necrolemur*, a small primate from the Eocene of Egypt (courtesy of Department of Geology, University of Saskatchewan).

than that, animals with stereo vision do not need to move their heads to see depth, which is useful because movement alerts prey and predators. Our vision has evolved with our brains, and it has been suggested that flying foxes and bats are related to primates, because their brains are similar.

The more complex of the primates are the anthropoid apes. These include the Old and New World monkeys, and also the gibbon and orang-utan, and the group that includes the gorilla, chimpanzee and people. The split between the Old and the New World primates seems to have taken place about 40 million years ago. Earlier, Africa and South America had separated, but were still relatively close where Brazil's north-east coast was slipping along the coast of West Africa, from Nigeria to Liberia. Strings of volcanic islands may have existed in the narrow sea between the two continents and the separating communities of life were probably linked by animals that hopped from island to island. Eventually the distance became too great to traverse. The South American monkeys seem to have evolved from an ancestor somewhat like *Tarsius*, as did the Old

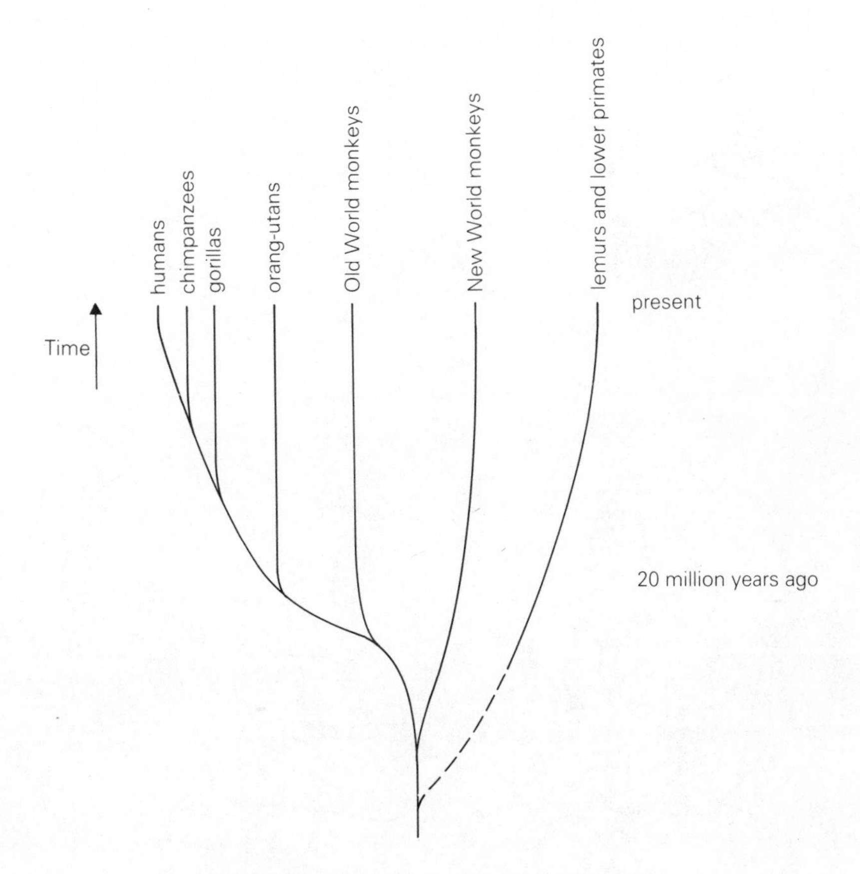

Figure 12.3 Humanity's place among the apes.

World line, but the two lines have been split ever since the Atlantic became uncrossable. The South American forms kept to the trees, and managed to develop their tails as a fifth hand; they are a very successful group, now threatened by their distant cousin, humanity.

In the Old World a second split took place in the primates, around 30 million years ago, between the line that led to the Old World Monkeys and the line of descent to the Great Apes, our own hominoid line. Around 20 million years ago, fossils record *Proconsul*, a baboon-sized animal that seems to have been a primitive, unspecialized hominoid, fruit-eating and with a difference in size between its sexes. The animal had elbow and shoulder joints and feet like a modern chimpanzee, a back like a gibbon, and a wrist like a monkey. It may have been close to the common ancestor of the large modern apes, including us.

Around this time, the lines that led to the modern gibbon and the orang-utan probably diverged from the African group. Also at about this time, the end of the early Miocene, the African plate was linked to the Eurasian plate: animals could migrate dry-footed across into Asia.

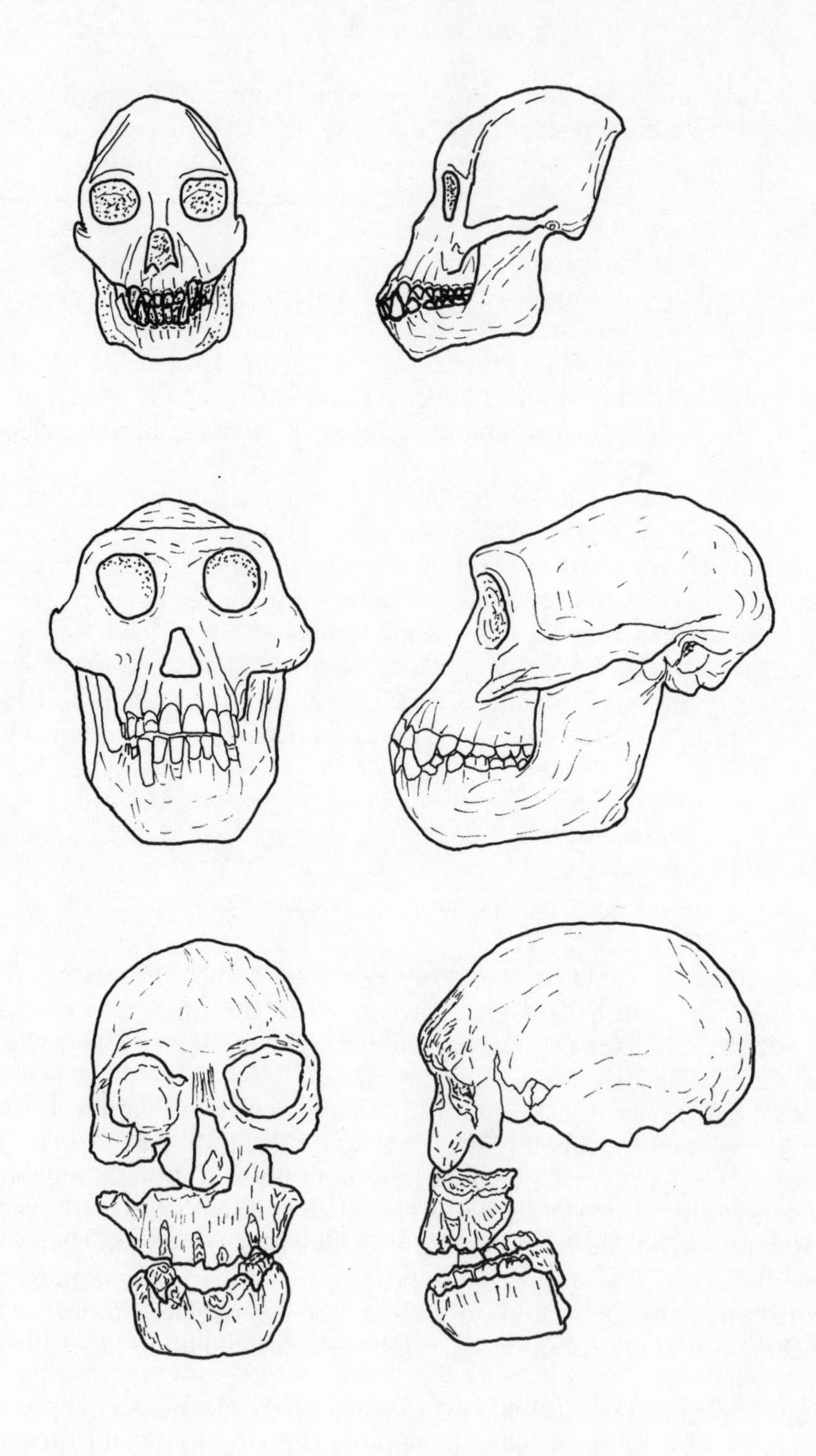

Figure 12.4 Ancestral relics. Top: skull of *Proconsul africanus*, a baboon-sized animal that lived in the early Miocene. Middle: *Australopithecus africanus*. Bottom: *Homo habilis*.

Between 7.5 and 4.5 million years ago, major climatic changes were taking place. About five million years ago, at the end of the Miocene, the Mediterranean became isolated from the Atlantic and dried up into a hot desert, scattered with salty lakes. The eventual breakthrough of the sea at Gibraltar may have been one of the great spectacles of the Earth's history, as the water flooded into the vast basin over cascades far greater than Niagara or the Victoria Falls. As David Livingstone would have put it, angels may have gazed on the scene in their flight, but our ancestors would not have appreciated it. In eastern Libya there is fossil evidence of a community which lived at the mouth of a large river draining into the Mediterranean. The fossils include fish, freshwater dolphins, crocodiles, turtles, snakes, monkeys, whales, hippos, elephants, and primitive hominids.

Our closest living relatives, gorillas and chimpanzees probably separated from our line (or us from them) about six to eight million years ago, according to molecular evidence. There is still much doubt about the relationships among the great apes. Humans, gorillas, common chimpanzees and pygmy chimpanzees, the four African species of great apes, seem to have had common ancestors, perhaps about ten million years ago or slightly before. One view is that a divergence took place then between the gorillas on one side and the three species of chimpanzees (common, pygmy and human) on the other side. This split was followed by a second divergence five to eight million years ago that divided the common and pygmy chimps from the third species of chimps, the humans. The evidence for this line of descent is molecular, but the results are still controversial.

THE EARLY HUMANS

About 3.6 million years ago at Laetoli in East Africa we have a record of human footprints, the first evidence of hominids walking upright. Our ancestor, roughly four million years ago, was an animal called *Australopithecus afarensis*. This animal (the famous skeleton of 'Lucy' is an example) probably weighed 25–50 kg (approx. 60–120 lbs) as an adult, the males much larger than the females. Its brain was probably about 400 cm^3 in volume. Their teeth seem to be adapted to fruits, seeds and roots, which needed much chewing. Their children matured slowly.

About two to three million years ago, there were several species of *Australopithecus*, including *Australopithecus afarensis* and, possibly derived through it (the exact sequence is controversial), *Australopithecus africanus* and the robust animals *Australopithecus boisei* and *Australopithecus robustus*. These robust species were probably ape-like vegetarians, chewing roots and tubers. Also descended from *Australopithecus afarensis* was another species, *Homo habilis* ('handyman'). This species was rather like *Australopithecus* in its face and teeth, but with a brain of about 700 cm^3 in volume, larger than the brains of its ancestors. The limb bones are also different, and the evidence seems to imply that *Homo habilis* ate more animal food. It may have been the first of our ancestors to use tools. Possibly there was a division of labour, the males hunting and the females gathering. The birth process also changed, with larger-headed infants. Human heads are so large that birth is very difficult, extraordinarily so when compared with, say, a hamster or with a chicken laying eggs. It has been suggested that human babies are born prematurely in comparison with other primates: the infant is not fully developed at birth and is initially helpless, because if a human baby were to be fully gestated by the standards of other animals it would have too large a head to pass through the birth canal. Helpless babies need constant care: in other words, they demand a social structure if the species is to survive.

Homo habilis existed for a few hundred thousand years. By about 1.6 million years ago, *Homo erectus* had appeared in Africa, and by one million years ago it lived also in south-east and east Asia. This species survived for over a million years, at least until 300 000 years ago. It had a larger brain (800 cm^3) than *Homo habilis*, and archaeological evidence suggests that it made stone tools such as hand axes. At some time during this period *Australopithecus* became extinct. Did it die out because of interspecies competition, or from drought and famine, or what? – we do not know, but it was our ancestors that survived, not our cousins. One of these beings, either *Homo habilis* or *Australopithecus robustus*, discovered the use of fire over a million years ago. The Swartkrans cave, in South Africa, contains burnt bones of animals and australopithecines. Some bones are butchered. Which of the hominids, *Homo* or *Australopithecus*, used fire first? We do not know. Possibly it was both; or perhaps our human ancestors destroyed the australopithecines.

HOMO SAPIENS

Homo erectus passed to the first *Homo sapiens*, including the Neanderthalers, and then to us. In a polygamous society, with rapid generation turnover, change can come quickly. A very few individuals can dominate the gene pool of the species. For instance, in seventeenth-century England, cattle were rapidly bred into the bewildering array on farms today, ranging from small dun hornless animals to monstrous multicoloured giants. Dogs, too, have been transformed from mongrels to the dachshund (for going down badger holes), the chihuahua and the great dane. They still interbreed (in principle at least, though a cross between a St Bernard and a chihuahua would tax the imagination), but their fossils will look quite different. Much of this evolution, especially in cattle, was done by selective breeding using males with particular characteristics. Polygamous humans presumably did the same.

Deliberate breeding of humans has, probably, been tried only once, when a King of Prussia decided to develop a new superweapon by breeding a regiment of giant soldiers. He commanded the largest men and women to procreate; this was done by the simple method of kidnapping tall men across Europe and putting them to stud with a collection of large girls. The Prussians even kidnapped the Abbé Bastiani while he was celebrating mass in an Italian church. The Abbé was indeed a giant, but Europe was not amused. As a means of producing large soldiers it would probably have worked (except that this was in the grand tradition of military bungles because, in a war fought with guns rather than swords, big soldiers are not much use, as they make excellent targets).

Neanderthal people had large brains – comparable in size to ours – and strong skeletons. They were probably fast-maturing: the fossil of a Neanderthal child appears more advanced than that of a modern human child of the same age. Perhaps they interbred with our ancestors to some extent, or possibly they were separate species: we do not know. In his poignant novel *The Inheritors*, William Golding explores their end as well as any palaeontologist can. In answer to the King of Prussia, he shows the value of every life, however unfavoured by evolution.

The transition to modern humans came sometime around and after 100 000 years ago, though we can look back somewhat further in our genes, to a common ancestor, or rather ancestress. The evidence is from human mitochondria. DNA sequences in the mitochondria are inherited from the mother only, not from the father. Study of human mitochondrial DNA suggests that we all share a common, distant ancestress, a many greats-grandmother, who lived perhaps 150 000 to 200 000 years ago, though possibly as long ago as a million years. The interpretation of the evidence is controversial, but it is possible that all living humans are descended from one woman. This does not necessarily mean that she was Eve (though she might have been): it is possible that one woman in a small tribe eventually became a common ancestress of all the members of the tribe, just as Queen Victoria was the common relative of most of Europe's royal families. In a tribe of a thousand people, with a constant population, after ten generations everyone will probably share some genes from nearly all of the mothers of the first generation. Modern humans seem to have begun with a single couple or a small group, probably in Africa. There is much debate about what happened next, but evidence from proteins can be interpreted as implying that a very small group – a few individuals or even a single couple – left Africa, possibly about 100 000 years ago, perhaps before. Remains of modern humans who lived 92 000 years ago have been found in Israel. From these modern humans came the billions who have filled the other continents.

The theological view of the origin of humanity is, of course, separate from and not dependent upon the biological evidence, despite the humorous naming of the mitochondrial ancestress as Eve. Theologically, humanity began with the first couple who had freedom of choice, who had souls. Neanderthalers buried their dead. They may have puzzled over their own existence and that of the world to come. An animal reacts to circumstances; it has no freedom, only stimulus and response. Humans have self-identity: they can say 'I'. Only a human can know right and wrong. Were the Neanderthalers human in this sense? Or did symbolic language, choice, and freedom to do evil come later? All that geologists can say is that modern humans, able to make sophisticated tools, to build homes, and to create art and language, are very new to the planet: they have been around for a few tens of thousands of years perhaps. Palaeontology and molecular biology have little business here, in the search for the origin of the soul.

The explosion came: in Europe about 30–40 000 years ago, the Neanderthalers were replaced, Asia was occupied, Australia was colonized. Perhaps 30 000 years ago North China was inhabited, and people moved north towards the Bering Crossing and Alaska. Exactly when humans entered the Americas is disputed – most of the evidence suggests about 15 000 years ago, but a few early possible tools have been found that suggest dates as long ago as 30 000 years. By 10 000 years ago, though, the whole planet had been colonized, except for a few islands such as Madagascar and New Zealand, which was occupied about a thousand years ago or less, and Antarctica. A single species had claimed the globe.

THE ICE AGES

During the Cretaceous most of the Earth's surface was warm. Solar heat energy that fell on tropical regions was efficiently transferred to the poles by the climate system. The early part of the Tertiary was a time of climatic fluctuations but, nevertheless, the climate remained mostly fairly warm. As late as the Eocene, forests grew in the farthest north of the Canadian Arctic. Alligators basked there in the midnight sun, and presumably slept in the polar night.

Change came, for reasons not yet understood. Perhaps the opening of the new oceans was important, causing carbon to be buried in the sediment and as gas and oil on continental margins. This would have caused the atmospheric inventory of carbon dioxide to fall. Perhaps the cause was the reduction in volcanism and a smaller supply of carbon dioxide to the biosphere. Possibly the slow changing of the Earth's orbit was important. Or perhaps the weather system changed as the continents drifted. We do not know.

The change began in the early Oligocene, when Antarctica became glaciated. By this time, 36 million years ago, ice sheets grew to cover most of that continent and to extend out to sea. However, the climate was warmer than today: the Antarctic ice sheets were probably temperate, not too far below freezing, in contrast to the very cold ice sheet that covers modern Antarctica. Heavy snowfall must have fed the ice; today Antarctica is a desert.

Not all of the continent was ice. A *Nothofagus* leaf, of the southern beech tree, has been found in a drill-core through 27 million-year-old sediments. Southern beech cannot cross the oceans, so the trees must have existed continuously until then on the fringes of the ice sheets, just as southern beech occurs today close to the glaciers of South Island, New Zealand. Multiple glaciation took place. Eventually trees became extinct on Antarctica.

More recently, four to five million years ago, Antarctica experienced deglaciation, and the Southern Ocean warmed. But about 2.4 million years ago, Antarctica was reglaciated, and the planet as a whole cooled markedly. There was fluctuation, but about 600 000 years ago intense glaciation set in and the ice sheets became very extensive.

Over the past half million years there have been about five cycles of planetary warming and cooling. In each cycle a gradual cooling takes place until full glacial conditions are reached, then sudden warming occurs and the ice quickly retreats. During the ice ages, 40–50 million cubic kilometres of ice build up on

Figure 12.5 *Mammuthus primigenius*, the woolly mammoth, about 3 m (10 ft) at the shoulder (courtesy of Department of Geology, University of Saskatchewan).

the land, making enormous ice domes in Antarctica, North America and western Eurasia. This was the setting in which our ancestors evolved. Part of the reason for the cyclic pattern of glaciation is now known: the shifts in temperature in the ice ages seem to correspond closely with subtle changes in the Earth's orbit and the tilt of its axis of rotation. This theory, proposed by M. Milankovitch, provides an excellent explanation of the cyclical ups and downs of temperature during the past half million or more years. In detail, however, the theory explains neither the suddenness nor the magnitude of some of the changes.

The past two glacial cycles are well recorded in the Antarctic ice. The ice has trapped bubbles of air, which can be recovered in drill-cores and analysed. From this, the carbon dioxide (CO_2) and methane (CH_4) contents of the ancient atmosphere, and its temperature, can be measured or calculated.

About 150 000 years ago, the Earth was deep in a glacial period. CO_2 levels in the air were roughly 190 parts per million (ppm); the temperature was cold. Suddenly, about 140 000 years ago, the system changed completely. CO_2 jumped to 280 ppm and the temperature rose. The transition was very rapid, as if the climate had crossed a threshold. During the warm times that followed, in pools near Cambridge, England, hippopotamuses bathed, while elephant-like animals looked on.

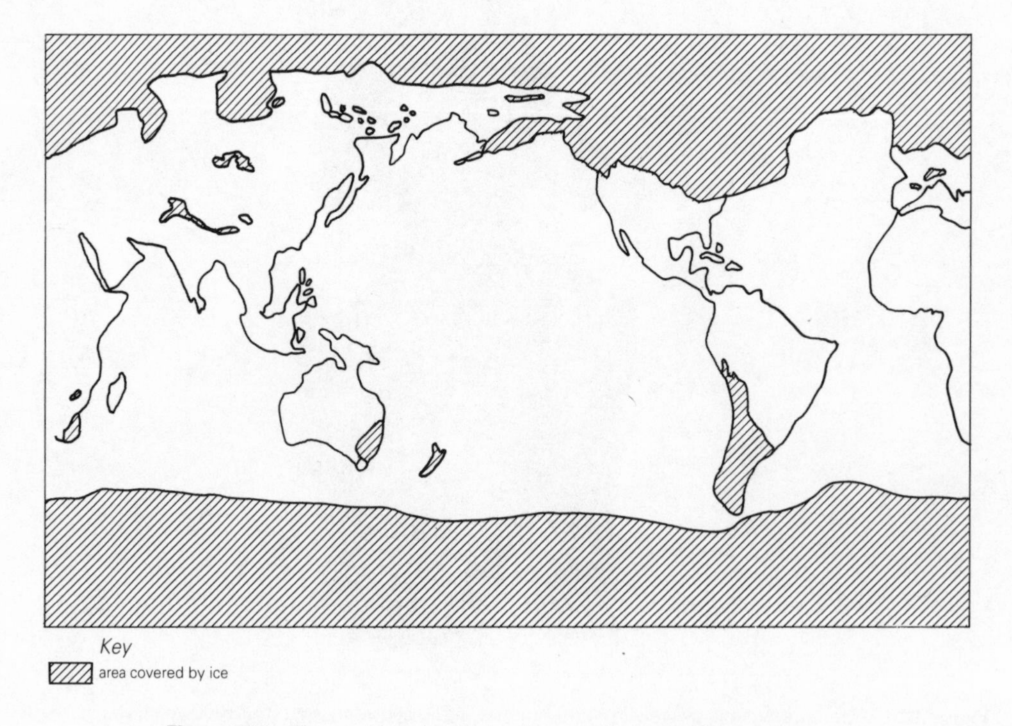

Figure 12.6 The world during the last ice age, about 18 000 years ago.

But the cold slowly returned and intensified. By 20 000 years ago the Earth was deep in a glacial event. Virtually all of Canada was covered by a huge ice dome, kilometres thick, centred around Hudson Bay. In Europe a similar dome over the Baltic spread out across Britain and Russia. The high latitudes were cold, dry and dusty. In the equatorial regions the seas remained warm, but worldwide the climate patterns were very different from today.

The end of the Ice Age came, as before, very suddenly. It happened about 13 500 years ago, around 11 500 BC. Almost instantaneously in geological terms, the methane (CH_4) and CO_2 content of the air jumped, and the temperature rose sharply. It took several thousand years to melt the giant ice domes of Canada and Scandinavia, but the climatic transition took only a few centuries or less. The climatic system crossed a threshold and took up a new, very different, stable state. The increase in CO_2 was about 70 ppm, from 190 to 260 ppm; the

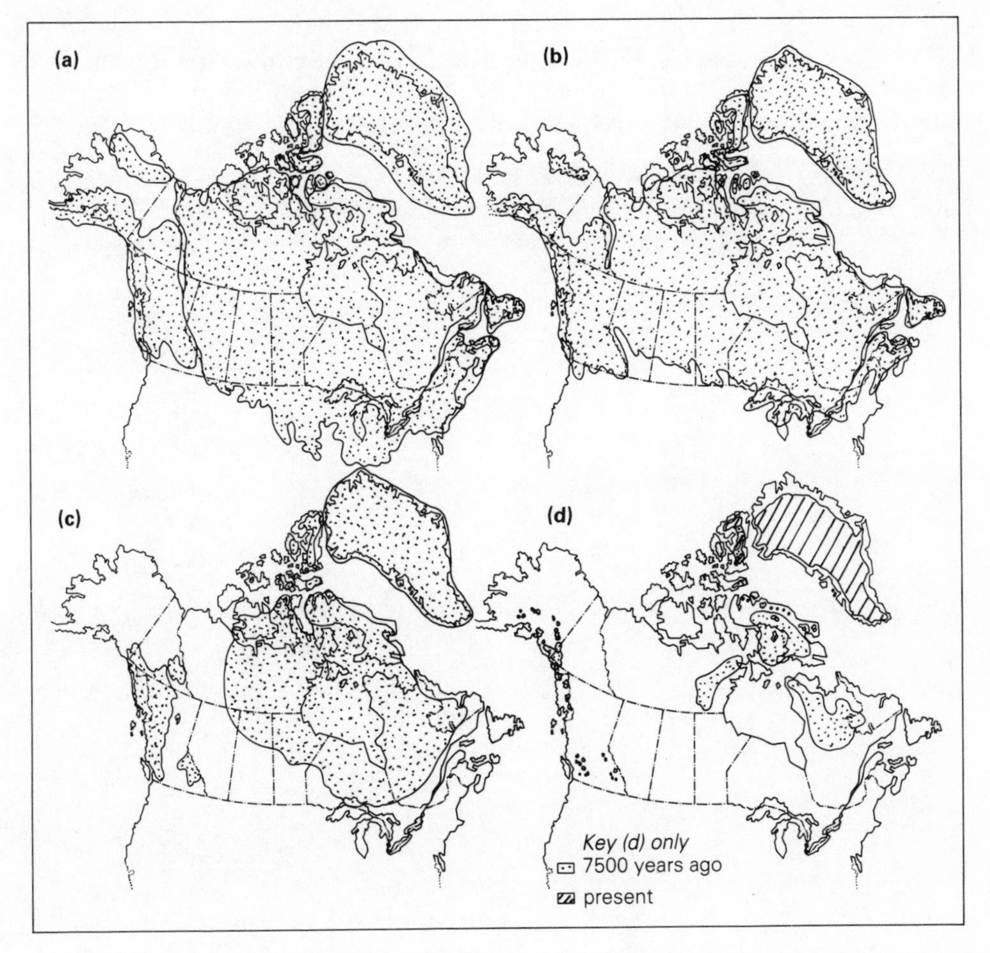

Figure 12.7 The retreat of the glaciers in North America. (a) 20 000 years ago; (b) 13 000 years ago; (c) 10 000 years ago; (d) 7500 years ago to present.

increase in CH_4 was about 320 parts per billion (ppb), from 330 to 650 ppb. This change, that seems to have ended the Ice Age, should be compared with the recent human-driven increase in the past century, from the post-glacial level of about 280 ppm CO_2 to about 350 ppm today, and of CH_4 to 1800 ppb, both still climbing rapidly.

The climatic change 13 500 years ago created our modern world. The ice slowly melted and was replaced by forest and tundra. Humanity moved north, across the Alps in Europe, and into Canada. The course of human history depends upon the change. We have now altered the air more, by our actions in this last century, than the alteration that drove the Earth out of glaciation. We do not know if another climatic threshold awaits us.

FURTHER READING

Brain, C.K. & A. Sillen 1988. Evidence from the Swartkrans cave for the earliest use of fire. *Nature* **336**, 464–6.

Reader, J. & J. Gurche 1986. *The rise of life: the first 3.5 billion years*. New York: A.A. Knopf/London: Roxby.

Simons, E.L. 1989. Human origins. *Science* **245**, 1343–50.

Walker, A. & M. Teaford 1989. *The hunt for* Proconsul. *Scientific American* **260**, 76–82.

13
Inhabiting an island

CONQUERING THE EARTH

The Ice Age ended and humanity claimed the planet. The changes to the planet that have occurred since then, in the past 10 000 years, are massive. In geological perspective, they are on a scale that is comparable to the greatest changes in the living community since the Cambrian. These changes will have permanent effect, though they have happened in a geological instant. They have been caused by the growth and actions of the human population. Human beings are not animals. They do not follow geological rules in their behaviour, and it is wrong to analyse their actions by simplistic analogues with earlier changes known from the fossil record. Nevertheless, we are subject still to the general constraints of climate, resources, and our own abilities, and it is worth looking at the effects of our rapid population growth both by studying the behaviour of animal communities that have undergone rapid growth in numbers, and by studying the human record in limited systems, such as islands. Conquerors often become tyrants.

The colonization of the continents by humans was associated with a wave of extinction of larger mammals. This wave of extinction has been alluded to in Chapter 10; it is worth discussing in more detail here. In Australia, North and South America, Madagascar and New Zealand the story was the same: the arrival of colonizing humans coincided with the sudden disappearance of most large animals. One wave of humans seems to have colonized Australia roughly 30 000 years ago. Their arrival was followed by the rapid extinction of nearly all the large marsupials – the rhino-like diprotodonts, the giant wombats, the ponderous giant kangaroos and the marsupial lions all went. By the time the colonial wave arrived from Europe, most of the big game in Australia had already been wiped out. Of the 22 genera of large animals living 30 000 years ago, only three genera remain.

In the Americas, the main human impact began after a wave of colonists arrived from Asia at the end of the last ice age, perhaps 14 000 years ago. Over the three million years prior to the last glaciation, 20 genera of large mammals

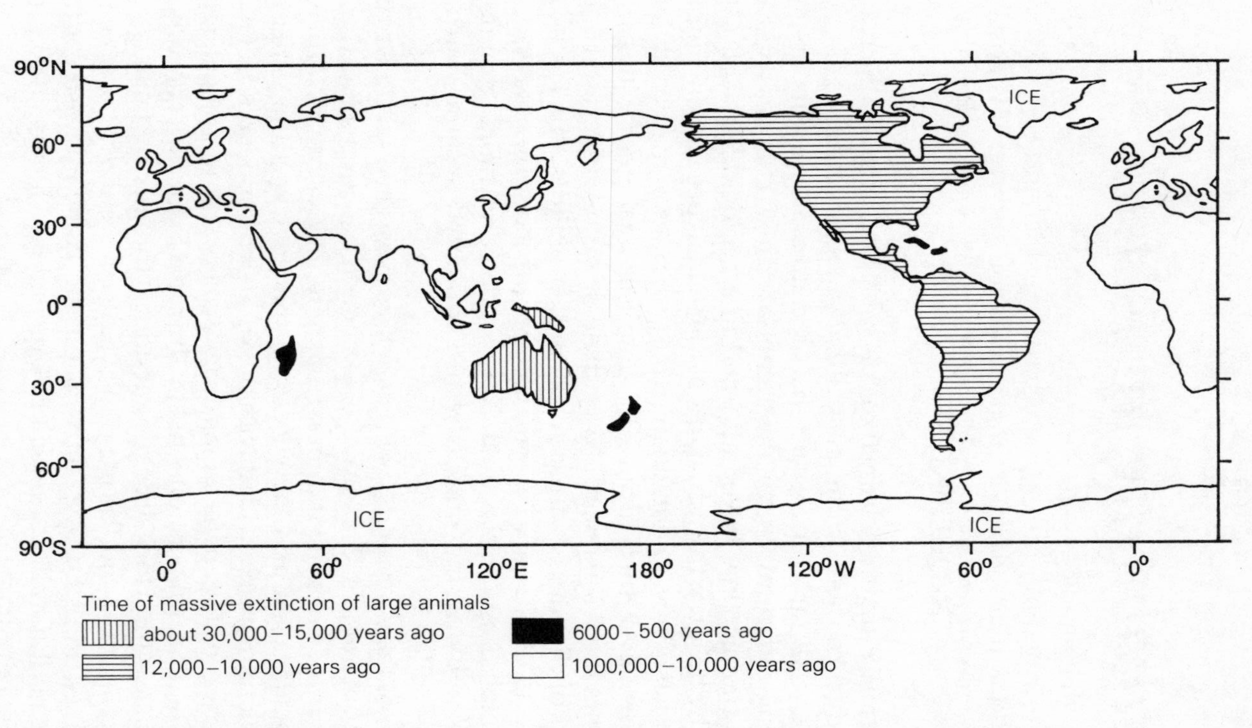

Figure 13.1 The timing of recent extinctions of large animals. In Africa, most large animals survived until recently, although some became extinct in the past 100 000 years. In Europe 37 genera of large mammals disappeared, such as the mammoth, woolly rhino and musk-ox, which became extinct 10–13 000 years ago. In Asia, the pattern was similar. Most Australian large animals became extinct after human colonization about 30 000 years ago. In the Americas, extinction was contemporary with human colonization 10–12 000 years ago.

had become extinct: this is the normal evolutionary turnover in times of climate change. In contrast, around 11 000 years ago, within a few centuries, 33 genera suddenly became extinct, most probably killed by humans. Twelve genera were left. In South America, 46 genera became extinct, and 12 were left. Mammoths, camels, horses, dire wolves, sabretooth cats, glyptodonts had all gone. The second colonial wave, from Europe, set about reducing the remainder, although the horse was reintroduced. In Madagascar and New Zealand the extinctions occurred later. The Maoris removed the moas of New Zealand, in part by deliberate burning of habitat, within a few centuries. The Madagascan people wiped out the elephant birds equally rapidly.

Only in Africa did a diverse large animal population survive. It is possible that these animals, having evolved with early humans, were better able to compete with them. It is also probable that in Africa humans had more enemies: insects, parasites and diseases that had specialized in attacking hominids. In Australia and the Americas these constraints on humans would not have been present, and the human population would have been limited only by the available food supply: the large animals, as long as they lasted.

In the Jewish Talmud there is a fascinating saying of Hillel: 'he that increases his flesh, increases vermin'. Imagine placing a pair of rats on an island like Big South Cape Island off New Zealand, teeming with assorted birds, trees, shrubs and insects, which over the millennia has come to attain a fluctuating but roughly balanced stability. When introduced to this system that has previously known no rats, the rats eat birds' eggs, and also the chicks. Rats have baby rats. Soon the birds are gone; the rats turn to the insects, the trees and the shrubs, eating pollinators and pollinated, fruit, seeds, and berries. The two original rats become millions.

Then, suddenly, all is gone. The island is devastated: the birds and insects have been eaten, the annual plants have gone after the destruction of a generation of seeds, the trees are dying, their bark stripped and seeds eaten. The rats turn upon themselves, cannibal. Eventually, stability re-emerges. The forest and birds are no more. The island is covered in wind-seeded grasses, with a few hardy weeds, and perhaps the odd seabird nesting on a cliff inaccessible to the rats. On the land, a few pairs of surviving rats scrabble for a living, eating insects and whatever seed they can find. Perhaps in a few million years a new richness will emerge: the rats will specialize into herbivores and carnivorous species. Vegetation will develop to match.

It is, of course, not only rats that do this: any competitive introduced species will devastate an old stability. Goats, pigs and cats are famous for this: Charles Darwin comments on the sad state of St Helena after their invasion. There are innumerable examples of similar irruptions on islands by introduced European domestic animals.

One of the most puzzling stories of population explosion and collapse comes from Easter Island, probably the most isolated inhabited land on Earth. The island is subtropical, and pollen cored from crater lakes shows that it formerly had a diverse forest vegetation, with abundant palm trees and a wide variety of shrubs. About a thousand years ago, people arrived. They developed an extraordinary culture, including the sculpting of the gigantic Easter Island statues in the period AD 1400–1680. Then, suddenly, everything collapsed: the first European visitors found an island mostly covered by grassland, the few inhabitants in a wretched state.

One possible reconstruction of what took place is that people acted exactly like rats – explosion, the flowering of a fine civilization, the destruction of most of the trees (now extinct on the island and surviving in botanical gardens). When the trees went, so did the opportunity to build wooden canoes to fish for food, and the food supply from the once fertile forest soils declined rapidly. Society collapsed in warfare. The eyes fell from the statues.

Disaster struck elsewhere in the Pacific islands too: explosion leading to collapse, and a few hardy survivors persisting in an impoverished culture now adapted to island life and culturally at ease with cannibalism. One legendary chief who became a fiercely loyal subject of Queen Victoria used to boast of the thousands he had eaten. Plump young girls tasted best: he gave them up on becoming a Methodist.

There seems to be a pattern. The rats show it; so do humans, if the Pacific legends are true. A greatly advantaged species arrives on an island, devours the abundant food supply, explodes in population, devastates its environment and collapses. Eventually an impoverished stability is restored, which may perhaps include a few members of the species that caused the disaster.

Whatever took place on Easter Island, the warning is sombre. The magnificent blind monoliths stare unmoving across a barren land and an empty ocean, a haunting premonition of the world that may come. Easter is also the time of Passover when Jewish and Christian believers contemplate the crossing over from an existence determined by the forces of this world – exploitation, prejudice, power – to a new unknown world. But whether that new world will be the world of the Sinai waste, a barren land in which a few struggle to survive, or whether it will be Zion depends on our choice. 'All is foreseen', says Rabbi Akiba in the Talmud, echoing St Paul, 'but freedom of choice is granted'.

THE RATS AT WORK

There is a another side to the Darwinian notion of natural selection and survival of the fittest: the hypothesis that all living organisms inevitably co-operate to create their environment. Our planet, like an island, is an intricately interconnected system: everything from bacterium to elephant shares in the management of the system. To displace one component, even slightly, means that the other components must adapt to the new setting. The system is enormously rich and flexible. It has weathered many challenges in the past; it is resilient. But the challenges we set it today are immense.

On land, there is an irruption of humanity. In a few thousand generations – a geological moment – we have rearranged the ecology of the continents. We have, in many cultures, an image of the 'noble savage' – that so-called aboriginal humans lived in harmony with the environment. The geological record says otherwise. From Australia to Zimbabwe humans have, from the beginning, made species extinct. Now, technological civilization has begun rearranging the vegetation. Large areas of Europe, India and China are now gardens. Nothing grows unless humans determine that it should grow, plants we do not want we fight as a weed. Central North America and the fertile fringe of Australia are close to being the same. In Africa and southern Asia we have devastated the tropical dry forest and a large part of the rainforest. The coastal forest of Brazil is almost gone. Everywhere grassland or farmland replaces forest, scrub replaces grassland, desert replaces scrub. In the equatorial belt we challenge two vital weather factories: the Amazon and the South-East Asian rain forests.

Weather is a subtle business. Small changes trigger great events. The global weather system is precariously balanced. If we remove its ecological pillars, while simultaneously altering the air by adding greenhouse gases, major changes may occur. For instance, the Amazon and Borneo rainforests are extremely important climatically; they are the thunderstorm factories of the world. Removal of the Amazon forest may markedly change the water circulation and heat movement in the equatorial atmosphere, perhaps shifting the pattern of the jet streams and the weather over North America. Around Borneo, the seas are warm, and storms would continue to occur, even without forest, but their style and frequency could well change, perhaps with far-reaching consequences for the people of India and China who depend on the monsoon.

The story of the oceans today is, too, the story of the Pacific islands: one species is destroying all others. The seas too have their forests, their man-made grasslands, their deserts. The catastrophe of Easter Island can happen there also. In the oceans, especially the Pacific, we slash and destroy through the patrimony of the aeons, decapitating complex food pyramids, cutting chaotically through the delicate fabric of nature. The great chain of being is broken.

The lesson of the world's islands and of the geological record tells us that success can lead to disaster. Our planet is not infinite. In *The tale without a head* (translated by Hazel Carter), one of the poets of Zimbabwe, Herbert Chitepo laments

'The water has dried up in the rivers
the fields are as bare as deserts
even the wild fruits
die in the belly of the Earth . . .

I stand here
looking behind and before
whence we have come is darkness
whither we go is darkness
Bring light, you elders
and guide us in this darkness.'

THE MANAGEMENT OF OUR EARTH

The Earth has been described as a 'spaceship'. This is not properly true. A spaceship provides a living environment where the atmosphere and temperature are mechanically and chemically controlled, for a short period, to allow a human being to live. This is a transient environment: it does not last. We could transform the planet into a spaceship, perhaps, with factories reprocessing the air fouled by factories, our Earth reduced to a place of humans and bacteria, the bacteria cultured in vast farms to sustain the air and feed the people. It would not be a very stable planet, however: spaceships, like submarines, must land and refuel.

An island is a better analogy: a place that has evolved an interactive community. The Earth is an island. The island community is balanced, so that it sustains its component life. Some things can be lost without catastrophe – we mourn the loss of the great auk and passenger pigeon, but their destruction did not destroy the economy of the planet, though it may have changed the ecology of the Northern Hemisphere. But the loss of each species that goes diminishes the life of the rest.

If the island community becomes too broken or too small, the security of the whole is imperilled. In New Zealand, 'the half-sunken ark', there are many endangered species of birds. Their habitat, naturally, is often protected. Even so, they often disappear if the area around them, which they do not appear to use, is destroyed. The reason is that their remaining 'islands' of habitat have become too constrained: the species become extinct. Most North American game parks probably suffer from this – too small to maintain the diversity of species to which they at present give refuge. Eventually, unless the parks are very carefully managed by man, some species will disappear until an impoverished stability is maintained. Only in the largest parks, such as in the Canadian Rockies, will the full North American diversity remain. In most of the land area of the rich nations, the habitat of wild nature has been broken up into small islets by uncrossable superhighways, that to lizards or hedgehogs are barriers as effective as an ocean. Within the remaining islets, species gradually become extinct from local causes, and there is no replenishment now possible from remaining populations elsewhere. In Africa, similar problems affect the large mammals. Zimbabwe has some of the best run game parks on Earth, but nevertheless elephant need to migrate. When I was a child, the sight of elephant moving across the country north of the Limpopo was an unappreciated familiar wonder. Today that land is given over to cattle – the elephant are confined to enormous but widely separated game parks in Hwange, the Zambesi valley, and Gona re Zhou. These parks should be enough, but they are not. Elephant population rises (elephants breed too), but the migration routes are closed. The only way to sustain the elephant population is to manage them, which means shooting them when the population rises beyond the long-term carrying capacity of their remaining land.

Recently, a group of ecologists at Stanford University and NASA estimated that human beings are appropriating a substantial fraction, perhaps up to two-fifths, of the biological productivity of the Earth. This estimate is controversial: some scientists have argued that it is too high, while others feel that it may understate our impact on the planet. Whatever the real proportion – and it increases each year – humans are now the most important biological factor in the ecology of the planet. Humanity now manages the planet.

We stand today on the threshold, on the edge of destroying ourselves and our home. Is our atmosphere to become what Hamlet called 'a foul and pestilent congregation of vapours', a dead thing above a noxious sea? Will we go like the dinosaurs? They died, leaving only the birds as a small memory of their ferocity. In his four last songs, written when Europe lay in ruins at the end of the last war, Richard Strauss closes an age of European history: the notes include the trill of a lark as a final memory of that terrible strength, now dead.

But there is hope too. The lark may be the survivor of the dinosaurs, but it is also one of the more gentle colonists of New Zealand. One of the delights of our world is the walk across Tongariro, a rugged volcano rising above the lakes and forests of the North Island: the only noise in all this beauty is the song of countless larks. They are imports from Britain, but they have achieved a synthesis into the complex new web of life that is slowly emerging there. The old Nature is dead, but it may be possible to create a substitute for the Garden of Eden.

There are many lessons for us in the history of life on Earth. We share, with all living beings, a common ancestry. Over geological time, our ancestors, together with the ancestors of the creatures that exist or have existed, have created complex communities of life. These communities have been built over millions of years, as interdependent webs of activity. Some of these communities have been very long-lasting, but on occasion, whole communities have disappeared. If there is too much perturbation of a system, the web of life may partly fail, only to be rebuilt again over the millions of years by other organisms. The changes that we have made in the past few centuries equal the greatest in the geological record: we are challenging the stability of the system that has given birth to us.

Today we live in a planet with no frontier. We are the first generation of humanity to encounter not boundless nature but a small closed island. Our future, surely, is to attempt to sustain what is left of our heritage, the life of our planet. If we can achieve that, we may then go beyond into space, to carry life there. If we fail, we may go the way of the dinosaurs. Whatever we do, we cannot deny our ability and sink back to stasis, for in stasis we die.

FURTHER READING

Friday, L. & R. Laskey (eds) 1989. *The fragile environment*. Cambridge: Cambridge University Press.

Goudie, A. 1990. *The human impact on the natural environment*, 3rd edn. Oxford: Blackwell.

King, C. 1984. *Immigrant killers: introduced predators and the conservation of birds in New Zealand*. Oxford: Oxford University Press.

Martin, P.S. & R.G. Klein 1984. *Quaternary extinctions: a prehistoric revolution*. Tucson, Ariz.: University of Arizona Press.

Mungall, C. & D.J. McLaren 1990. *Planet under stress*. Toronto: Oxford University Press.

Nisbet, E.G. 1991. *Leaving Eden*. Cambridge: Cambridge University Press.

Tickell, C. 1986. *Climatic change and world affairs*. Lanham, Md: University Press of America.

Index

Figure numbers are shown in italic.